Synthese Library

Studies in Epistemology, Logic, Methodology, and Philosophy of Science

Volume 430

The aim of *Synthese Library* is to provide a forum for the best current work in the methodology and philosophy of science and in epistemology. A wide variety of different approaches have traditionally been represented in the Library, and every effort is made to maintain this variety, not for its own sake, but because we believe that there are many fruitful and illuminating approaches to the philosophy of science and related disciplines.

Special attention is paid to methodological studies which illustrate the interplay of empirical and philosophical viewpoints and to contributions to the formal (logical, set-theoretical, mathematical, information-theoretical, decision-theoretical, etc.) methodology of empirical sciences. Likewise, the applications of logical methods to epistemology as well as philosophically and methodologically relevant studies in logic are strongly encouraged. The emphasis on logic will be tempered by interest in the psychological, historical, and sociological aspects of science.

Besides monographs *Synthese Library* publishes thematically unified anthologies and edited volumes with a well-defined topical focus inside the aim and scope of the book series. The contributions in the volumes are expected to be focused and structurally organized in accordance with the central theme(s), and should be tied together by an extensive editorial introduction or set of introductions if the volume is divided into parts. An extensive bibliography and index are mandatory.

More information about this series at http://www.springer.com/series/6607

Wenceslao J. Gonzalez

Editor

Methodological Prospects for Scientific Research

From Pragmatism to Pluralism

 Springer

Editor
Wenceslao J. Gonzalez
Faculty of Humanities and Information
Science
University of A Coruña
Ferrol, Spain

ISSN 0166-6991   ISSN 2542-8292   (electronic)
Synthese Library
ISBN 978-3-030-52502-6   ISBN 978-3-030-52500-2   (eBook)
https://doi.org/10.1007/978-3-030-52500-2

This Springer imprint is published by the registered company Springer Nature Switzerland AG.
The registered company address is: Gewerbestrasse 11, 6330 Cham, Switzerland

Contents

Chapter 1
Pragmatism and Pluralism as Methodological Alternatives to Monism, Reductionism and Universalism

Wenceslao J. Gonzalez

Abstract Both pragmatism and pluralism are considered as methodological alternatives to monism, reductionism and universalism, and also to methodological imperialism. First, a historical framework in Suppes ([1978] 1981) provides a precedent for the current emphasis on methodological pluralism, while methodological pragmatism, which is explicit from Rescher (1977), has often had supporters, with greater or lesser intensity and conceived from different epistemological angles. From the 1980's to the initial years of this century, there has been a renewed interest in methodological pragmatism and pluralism, particularly since 2006. Second, there is a thematic view, which is focused on the central tenets of pragmatism and pluralism to outline their consequences for scientific research as alternative to the influential conceptions mentioned. The emphasis on practice, which leads to efficacy in realizing the collective goals of research, and the recognition of the diversity in scientific research, which is in tune with the polyhedral character of reality (natural, social and artificial) and its ontological levels (micro, meso, macro), contribute to the methodology of science in its twofold condition of analytical and prescriptive (in the sense of indicative patterns). Third, the structure and origin of this book is presented in connection with the previous steps.

Keywords Pragmatism · Pluralism · Methodological · Alternatives · Monism · Reductionism · Universalism

I am grateful to Nicholas Rescher for his remarks on this paper, the final version of which I prepared at the Centre for Philosophy of Natural and Social Sciences (London School of Economics).

W. J. Gonzalez (✉)
Center for Research in Philosophy of Science and Technology, Faculty of Humanities and Information Science, University of A Coruña, Campus of Ferrol, Ferrol, Spain
e-mail: wenceslao.gonzalez@udc.es

Pragmatism and pluralism offer methodological alternatives to monism, reductionism and methodological universalism, and also to methodological imperialism.[1] In this regard, the emphasis on the progress of science as a *human activity*, rather than a mere content oriented to a cognitive goal,[2] and the inquiry made according to the kind of object and the *scale of reality* studied (micro, meso and macro) can support the advantages of pragmatism and pluralism as philosophico-methodological prospects for scientific research.[3] In addition, the existence of complex systems in the world (natural, social and artificial) (Mainzer [1994] 2007; Mitchell 2009), both structural and dynamic (Bertuglia and Vaio 2005), can play a role in favor of methodological approaches based on pragmatism or pluralism (or even a combination of both).

These advantages in the characterization of scientific advancements include the twofold condition of methodology of science as *analytical* and *prescriptive* (in the sense of indicative patterns instead of normative), insofar as (i) it analyzes how science *is* developed de facto, either in the research made in the past or in the present, and (ii) it proposes how it is possible to improve scientific research, based on how we think science *ought to be* developed in order to achieve its aims (in the short, middle or long run), according to the level of reality studied. Thus, the attention to human activity, the scale of reality and the structural and dynamic complexity can contribute to central features of the procedures and methods used in science.[4]

Both pragmatism and pluralism are, in principle, open to a *diversity of methods* in science, in general, in a group of sciences (natural, social or the artificial),[5] and in specific sciences (physics, economics, computer sciences, etc.). This implies that they do not start from a macrotheoretical scheme of unity of science or from the need for a methodological unification of sciences, achieved through an effort of convergence (commonly based on logical, epistemological or ontological criteria).

In addition, they normally consider the *practice* of science as a relevant factor, rather than just focusing on theoretical contributions to science, and they are generally sensitive to the importance of contexts (social, cultural, economic, etc.) for the advancement of science. Thus, to a large extent, these methodological approaches

---

[1]Besides the methodological component, pragmatism and pluralism can be considered in the semantic, logic, epistemological, ontological, axiological and ethical aspects of science.

[2]Pragmatism regarding research can take many forms, such as scientific progress focused on "the ways in which inquiry is adapted to meeting human needs," Kitcher (2015, 475). This view on methodological pragmatism is compatible with scientific realism, cf. Kitcher (2011b, 171–189).

[3]Perspectivism can be seen as a form of pluralism, mainly from an epistemological viewpoint, cf. Giere (2006a). Ronald Giere has explored "the extent to which a perspectival understanding of scientific knowledge supports forms of scientific pluralism," Giere (2006b, p. 26). But the analysis made here is broader than that view. Furthermore, this paper is focused on the methodological approach, which has a connection with an ontological basis and includes the development of the epistemological component.

[4]The distinction between "procedures" and "methods," which has a general character, in the case of scientific prediction is particularly clear, cf. Gonzalez (2015a, 255–273). See also Gonzalez (2020c).

[5]The sciences of design are included in the third group, cf. Simon (1996).

can overcome the key problems of monism,[6] reductionism[7] and methodological universalism as well as that of methodological imperialism.[8]

1.1   A Historical Framework: From Pioneers to Recent Approaches

Regarding the trajectory followed by pragmatism and pluralism as methodological alternatives to monism, reductionism and universalism, we can distinguish three moments. The first corresponds to the pioneers, Patrick Suppes and Nicholas Rescher, who open the methodological debate on pluralism and pragmatism respectively. The second is a series of authors who, from then on, have proposed theses of methodological pluralism or methodological pragmatism, which coexist with other conceptions that, since 1980, have been particularly influential. The third moment, which can be situated in 2006, implies a greater presence of pluralism and pragmatism in the methodological debates.

From a historical point of view, Patrick Suppes can be considered a precedent of the current methodological pluralism. In his famous text "The Plurality of Science" delivered at the Philosophy of Science Association meeting in 1978, he criticizes the idea of unity of science by Otto Neurath and defends the plurality of science (Suppes [1978] 1981). In this respect, Suppes questions the reductionism of language, the reductionism of the subject-matter and reductionism of method (Suppes [1978] 1981, 5–9). He does not find a methodological unity in science but rather diversity. Almost three decades later, in 1996, Ian Hacking — who was co-editor of the volume in which Suppes' text was published — raises the issue from the perspective of "The Disunities of the Sciences" (Hacking 1996).

But Hacking, who highlights the role of the history of science for philosophy of science, approaches the topic (unity vs. disunity, unification vs. plurality) from the perspective of the *sciences* — plural — rather than science (singular), as Suppes does. In this regard, Hacking sees precedents of scientific pluralism in terms of disunity between sciences in the nineteenth century, following diverse styles in doing science. He looks at the scientific differences pointed out by William Whewell and August Comte, two thinkers who had clearly different philosophico-methodological approaches (Hacking 1996, 37–39).

---

[6]Some of the problems can be seen in the trajectory followed by the methodological conceptions of logical positivism, logical empiricism and the received view, which were focused on verification, verifiability and empirical confirmation, cf. Suppe ([1974] 1977).

[7]"One form or another of reductionism has been central to the discussion of unity of science for a very long time," Suppes ([1978] 1981, p. 5).

[8]On the characterization of methodological universalism and methodological imperialism, see Gonzalez (2012).

Over the past few decades, the issue of pluralism has been present in philosophy and methodology of science from two main angles. First, from a general point of view, commonly according to four main approaches: naturalism (Dupré 1993), the social turn in science (Longino 1990),[9] scientific realism (Cartwright 2004; Psillos 2009),[10] and conceptions based on probability theories (Galavotti 2008). Second, in the analysis of positions within various sciences, be they disciplines investigating nature, such as biology (Mitchell 1992, 1993), or researching society, as in the case of economics (van Bouwel 2004). These analyses show a combination of novelty and continuity in philosophy and methodology of science.

Meanwhile, a pioneer in the approach to methodological pragmatism is Nicholas Rescher, who published a book on the subject in 1977. Since his volume (Rescher 1977), methodological pragmatism — with greater or lesser intensity and conceived from different epistemological angles — [11] has often had supporters. In fact, approaches for methodological pragmatism can be found in conceptions situated epistemologically in the four general lines indicated: naturalist thinkers,[12] supporters of the emphasis on the social dimension of science (Kitcher 2011a; cf. Gonzalez 2011), philosophers prone to scientific realism[13] and Bayesians of various tendencies (Gelman 2011; Bergman 2009). In addition, authors together with authors of a more personal stamp, such as Rescher, who develops a conceptually based pragmatic idealism (Rescher 1973, 1992).

After these pioneers — since the 1980s — some general philosophico-methodological approaches were particularly influential: (1) the "naturalist turn," which includes a cognitive characterization of science, based on cognitive sciences, the "normative naturalism," developed upon an interpretation of history of science, and other kinds of naturalisms centered on other sciences (mainly biology); (2) the "social turn," where the social concern for science puts the emphasis on the external perspective of scientific activity and the relations between science, technology and society; (3) the interest in "scientific realism," which led to a plethora of realist conceptions on science, because it can follow many lines (semantic, logical, epistemological, methodological, ontological, axiological, . . .), where the structural realism initiated by John Worrall and Philip Kitcher's analysis of the advancement of science have an important role; and (4) the emphasis on

---

[9]The social turn in science is also dominant in the book edited by Galison and Stump (1996).

[10]Also, from a critical point of view, see Chang (2012).

[11]These epistemological differences can be seen in the basis of the initial conceptions of American pragmatism. Rescher drew attention to these differences and has opted for Charles S. Peirce. See Rescher (2012b).

[12]In the problem-solving of naturalist kind, there can be pragmatic components of an internal character — in the heuristic processes — and of an external nature (how to do the research with collaborators), including the sciences of the artificial as empirical sciences. See Simon, "The Scientist as a Problem-Solver," in Simon (1991, 368–387).

[13]"In my interpretation, Peirce accepted the basic idea of the correspondence theory, but wanted to find a coextensive characterization of this concept." Niiniluoto (1999, p. 101).

the probabilistic viewpoint of science, where the Bayesian approach is discussed intensely.[14]

During recent years, these general philosophico-methodological approaches to science have continued to be influential and very much alive, such as the scientific realism debate, and some of them are compatible with versions of pragmatism and pluralism (Gonzalez 2020a). But there is now a *renewed* philosophico-methodological interest in pragmatism and pluralism, especially conceived as alternatives to monism, reductionism and universalism as well as to methodological imperialism. This is the starting point of this book on methodological prospects for scientific research.[15]

Besides the renewed interest in methodological pragmatism in science, either *per se* — focused on the importance of science as a practice, diversified according to the goals sought, and its effectiveness in achieving results — or in connection with other views, such as scientific realism (Gonzalez 2020b), there has been an increasing attention to methodological pluralism in modes related to science itself, groups of sciences and specific sciences. This is the case at least since the publication in 2006 of the book *Scientific Pluralism* in the Minnesota Studies in the Philosophy of Science (Kellert et al. 2006a).

Thereafter, this topic has become a trend for new analyses in favor of a polyhedral vision of scientific research: (a) As regards *science itself*, these views commonly think of the problem of the unity of science in a different way from the past.[16] This view is normally critical of methodological universalism and emphasizes elements of diversity in scientific research in empirical sciences instead of seeking a clear-cut monism (i.e., a single, complete and comprehensive scheme of the world, as a whole, or of a given realm).[17] (b) Concerning the *group of sciences*, there is a clear interest in interdisciplinarity,[18] multidisciplinarity, transdisciplinarity and crossdisciplinarity. This is based on a methodological view open to novelty, which is different from traditional forms of reductionism and in favor of cooperation and complementarity among disciplines. (c) Regarding *specific sciences*, methodological pluralism is in favor of taking into account the idiosyncrasies of the disciplines

---

[14]These conceptions with their bibliographical contributions can be found in Gonzalez (2006).

[15]This is a broader discussion than the methodological controversy between natural sciences and social sciences around the distinction *Erklären-Verstehen*, where there are at least nine options, cf. Gonzalez (2015b, 173–179) available in: https://doi.org/10.1080/02698595.2015.1119418 (accessed on 25.1.2017).

[16]Nevertheless, some authors have "explored an alternative vision of the 'Unity of Science' offered by the work of Vienna Circle cofounder Otto Neurath, which sees an irreducible variety of scientific disciplines cooperating for concrete purposes" Kellert et al. (2006b, p. vii). In this vein, see Richardson (2006).

[17]On the characterization of monism, see Kellert et al. (2006b, x).

[18]"Interdisciplinarity is one of the most acute challenges of contemporary science and university policy. It is widely believed that the most fertile novel ideas are born in the interface areas of various disciplinary approaches. The progress of science – including natural, human, and social sciences – thus requires the creation of research environments which foster interaction between scholars from different fields" (Niiniluoto 2020, 231).

rather than having a direct interest in a methodological imperialism (Gonzalez 2007).

1.2 A Thematic View: The "Internal" Side and the "External" Facet

Certainly, from a thematic viewpoint, methodological pluralism cannot be merely the existence of a plurality of sciences, to which Hacking refers when he mentions nineteenth century authors such as Whewell and Comte (Hacking 1996, 37–39), since these thinkers really thought of scientific methods that could be used in the various sciences. From the methodological point of view, they did not think of highlighting a diversity in scientific research but rather of what was common or shared by various sciences when investigating nature or society.

However, there is a diversity of objects of study, procedures and research methods, and solutions to the problems raised, which depend on the questions put forward and the context in which they are posed. Methodological pluralism adds more elements, above all because of its connection with an epistemological pluralism in posing the problems and an ontological diversity in terms of the diversity of levels of reality and the very polyhedral texture of the real. Thus, it is not a mere de facto "plurality of sciences," but rather a *plurality of science* in its different methodological expressions: formal and empirical; basic, applied and of application; natural, social and artificial; and within each science, such as pharmacology (as ongoing research on Covid-19 shows every day).

On the methodological side, pragmatism "pivots on the idea that the resolution of philosophical problems and issues should be developed in the light of consideration of effectiveness, efficiency, and success in matters of goal realization" (Rescher 2019, 58). Within the field of science this — as Rescher insists — does not mean instrumentalism, since it does not rest on criteria of utility but on the search for truth within the practical sector. Thus, from a methodological point of view, "successful praxis is seen as adequacy indicative with respect to our generalized criteria of acceptance and verification. The link between truth and praxis is not seen as direct, but as mediated by the methodology of inquiry. Our justificatory rule is: Truth claims via methods and methods via praxis" (Rescher 2019, 83).

1.2.1 Main Methodological Elements Involved

If we use the concept of "consensus" in its two main meanings — in a substantive sense and in its procedural dimension — [19] we then have the following. (a) Methodological pluralism implies that there is no methodological consensus in the *substantive sense*, insofar as it maintains that a universal or universalizable scientific method is not suitable, either now or in the possible future. (b) With respect to the methodological pluralism of consensus in the *procedural dimension*, which considers that we can come to agree on a series of common or shared features among all research methods, there are nuances according to the different orientations of methodological pluralism. For a scientific realist open to methodological pluralism, we can think of it in terms of *in pluribus unum*; while for a methodological pluralist who relies on relativist conceptions (for example, of Kuhnian inspiration within the "social turn" in philosophy of science) it may be something unviable.

Thematically, methodological pragmatism is focused on research processes that have to be effective and, if possible, efficient. Thus, as Nicholas Rescher states in his paper, "pragmatism adopts the procedure of assessing process by product, of evaluating ways of doing things on the basis of their functional efficacy in realizing the collective objectives. This can be done either *directly* by assessing the quality of the product, or *obliquely* by assessing the efficacy of procedures and methods in engendering high quality products. Methodological (as contrasted with Productive) Pragmatism adopts the latter approach" (Rescher 2020, 69). In this regard, he considers that "its prime advantage is its greater reliability in providing for quality control thanks to the superior realism of the evaluations it underwrites" (Rescher 2020, 69).

Following the recent trajectory in the methodological prospects to science, there are two main features involved: (I) an "internal" side, which is directly related to the scientific progress seen in pragmatist and pluralist perspectives, and (II) an "external" facet, which is connected to at least three kinds of social mediations: (i) the scientists as researchers that interact with other colleagues, (ii) the organizations where they develop basic science, applied science or application of science,[20] and (iii) the public policy on scientific research (international, national or regional) due to the guidelines and economic support for research, development and innovation $(R + D + i)$.[21]

---

[19]The notion of "consensus" has been considered by Rescher, who has defended pluralism against a "dogmatic uniformitarianism" and a "relativistic indifferentism." See Rescher (1993).

[20]The distinction between applied science and application of science is suggested by Ilkka Niiniluoto (1993, 1–21; especially, 9 and 19). A development of this methodological distinction is in Gonzalez (2013a) and Gonzalez (2015a, 2, 4, 12, 18, 31–34, 37–40, 70–71, 149–151, 317–321, 325, 330 and 335).

[21]Scientific research during the crisis generated by Covid-19 has clearly exemplified the three kinds of social mediation.

Concerning the "internal" side, pragmatist and pluralist perspectives need to be better off than monism, reductionism and universalism in order to tackle the levels of reality (micro, meso and macro) according to the diverse angles of research, which are those related to the aims of the inquiry, the kind of processes to be used and the expected results of this human activity. This includes considering the limits of the methods used. In this regard, Nicholas Rescher, a well-known pragmatic idealist,[22] criticizes the possibility of a pragmatic completeness of science: "could we ever be in a position to claim that science has been completed on the basis of success of its practical applications?" (Rescher 2012a, 156).

Methodological pragmatism highlights the primacy of the *practice* based on the activity of our science. It looks at the results obtained and the efficacy in realizing the *collective goals* of the scientific research (cf. Rescher 2020). Meanwhile, methodological pluralism emphasizes *diversity*. Thus, following the analytical condition of the methodology of science, it accepts the existence of multiple approaches in scientific research, each revealing different sides of a phenomenon. Thus, "there can be plurality of representational or classificatory schemes, of explanatory strategies, of models and theories, and of investigative questions and the strategies appropriate for answering them" (Kellert et al. 2006b, ix).

Regarding the "external" facet, what is involved is, at least, decision making by the researchers, the organizations where they work and the entities responsible of the public policy. In one way or another, they need to deal with methodological conceptions that have been very influential over the years. De facto, monism based on a key science (physics, biology, neuroscience, . . .) has received a lot of attention in research projects, which have often assumed the legacy — implicit or explicit — of the ideal of "unified science."

It happens that some version of monism appears very frequently in the presentation of science in terms of "scientism" through social media, either of a general character or focused just on the primacy of one kind of scientific approach (physicalist, biologist, neuroscientist, etc.). Furthermore, reductionism — both implicit and explicit — is still in place in textbooks. This is also the case of methodological universalism, which has a number of options (Gonzalez 2012, 156–162), and which is commonly accepted each time that "the" scientific method is presented.[23]

According to pragmatism and pluralism as methodological alternatives to monism, reductionism and methodological universalism, there is a large number of options, due to the *scope* to deal with the objects to be studied (processes, actions, etc.), the characteristics of the *methods* to be used to address the problems (basic, applied or the application) and the *results* that can be expected (in the short, middle

---

[22] Among his works on pragmatism in recent years are Rescher (2012b, 2014).

[23] John Worrall has made it explicit Karl Popper did not believe in something like "the" scientific method, conceived as a systematic way to achieve well founded results (Worrall 2001, 114). "As a rule, I begin my lectures on Scientific Method by telling my students that scientific method does not exist" (Popper [1956], 1992, 5).

or long run). These pragmatic and pluralist approaches can have a wide spectrum of possibilities, because they can have a quite different level in the intensity of the view held, such as modest and radical interpretations[24] or weak and strong conceptions.[25]

Commonly, these characterizations depend on the distance from the methodological relativism and the extent of their openness to objectivity in scientific research. Furthermore, if objectivity is accepted in the methodological approaches, then there is a variety of possibilities for scientific progress. These include a number of consequences for basic science, applied science and application of science as well as for the research made according to the scales of reality (micro, meso and macro) and the kind of reality involved (natural, social or artificial). Among them are some key methodological aspects.

1.2.2   Consequences of Pluralism and Pragmatism: Some Key Methodological Aspects

Based on the contributions made by pluralism and pragmatism within the thematic perspective, a number of particularly relevant methodological consequences can be highlighted. They deal with the issues raised by monism, reductionism and methodological universalism. They seek to reflect how research processes *actually are* and how they *ought to be*. Thus, they address the *analytical aspect*, which studies the procedures and methods used in scientific research, and the *prescriptive component*, which considers how to improve the way in which scientific knowledge is advanced.

(i) The monist aspiration to a *single and full account* of reality — at least of nature — is increasingly more difficult for a number of reasons. These include, firstly, the development of new sciences in the study of nature and society, but above all in the artificial realm (such as the sciences of the Internet) (Gonzalez and Arrojo 2019). They raise new objects of study, hitherto unsolved problems, and new methods in line with these problems, as is the case of design sciences (Gonzalez 2008, 2017). Secondly, within each discipline there is the principle of proliferation of questions regarding the reality researched, where every solution given to a problem raises a new question, which has to be answered.[26]

As a consequence of the widening of the field of study and the deepening of the areas already known, comes the novelty. In some cases, such as the sciences

---

[24]This is the case of pluralist interpretations, cf. Kellert et al. (2006b, xi–xiii)

[25]"Philosophers who advocate pluralism can and do differ as to the extent of the plurality they attribute to the sciences, the strength of the pluralism they adapt, and the broader philosophical implications they draw from it" (Kellert et al. 2006b, p. x)

[26]This principle is an issue that Rescher insists on in the context of the future knowledge. Cf. Rescher (2012a).

of design related to the Internet (Tiropanis et al. 2015), the novelty is vertical or transversal. Meanwhile, in other cases, there might be a horizontal or longitudinal novelty, insofar as there is a structural similarity in the phenomena studied. Both types of novelty highlight the dynamic complexity of the phenomena and, consequently, the historicity of science (cf. Gonzalez 2018).

(ii) Any attempt *to reduce* one science into another one can involve certain epistemological, methodological and ontological costs.[27] A methodological reduction can, in principle, be thought of in several ways, among which there are three: (a) as a simplification of theoretical contents, due to certain procedures and methods, in favor of a common ground of some scientific theories, (b) as a type of methodological subsumption, where some elements in one scientific level (micro, meso, macro) can be integrated in another level of knowledge or reality, and (c) as "the explanation of a theory or a set of experimental laws established in one area of inquiry, by a theory usually though not invariably formulated for some other domain" (Nagel 1961, 338).

But there are clear problems with methodological reductionism. Thus, besides the issue of the *emergency* of properties within a given system,[28] an important factor is the existence of *complexity*, both structural and dynamic.[29] This often leads to new aspects in the sciences already known and even to the development of new sciences to deal with the new objects, the novel problems and the methods to achieve the solutions. This methodological novelty is also the case when the complex systems are hierarchical and capable of near-decomposability.[30] Furthermore, it seems clear that some sciences have been commonly neglected in the philosophical analysis of science, and the philosophico-methodological results are often obtained in the analyses made in few sciences (mainly, natural sciences).

Historically, both in terms of the analytical aspect of methodology of science and regarding its prescriptive component, it seems clear that there has been a more intense level of philosophico-methodological attention to some sciences (physics, biology, . . .) and to the group of natural sciences in comparison with the group of social sciences. Meanwhile, the sciences of the artificial have received little overall attention. This trajectory could be for several reasons, including a combination of structural elements and dynamic components. Among these reasons, there are three *structural factors* that can be emphasized.

First, the existence of a well-established field conceived as a model for other sciences, such as physics for the group of natural sciences and even for science, in general. Second, the development of a domain having many epistemological

---

[27] On the issue of the "formal" and "nonformal" conditions for reductions in science, see the influential analysis made in Nagel (1961, 354–366). This double set of conditions can be used to consider the philosophico-methodological costs of the reduction.

[28] An analysis that highlights the *dynamic aspect* of the emergency, as opposed to the traditional emphasis on the structural facet of the emergency, is found in Humphreys (2016).

[29] See, for example, the case of economics, Rosser ([1999], 2004).

[30] One of the contributions of Herbert Simon was in the configuration of studies on complex systems, cf. Simon (1977, 1999, 2001).

and methodological influences on other scientific subjects, such as biology (mainly through the influential Darwinian approach). Third, the presence of a scientific terrain having a lot of practical consequences, especially by dint of being an applied science (such as economics, psychology or computer sciences) (Gonzalez 2013b, p. 1).

Nevertheless, there are other reasons that include clear *dynamic traits*. These can contribute to understanding the preferences among philosophers of science in favor of some disciplines rather than others as well as preferences towards one group of sciences instead of others. Among such reasons three can be recalled: "(a) the novelty of the scientific field, which makes the philosophical-methodological analysis more difficult; (b) the scarcity of explicit influences of the science beyond the 'boundaries' of the field; and (c) the strong interweaving with technology, which makes the distinction between the scientific approach and the technological constituent particularly difficult" (Gonzalez 2013b, 1).

(iii) Increasingly, there are more problems for methodological universalism and for the proposals being made of methodological imperialism. De facto, the search for a *universal method*, valid in principle for all the sciences, or a *methodological imperialism* (of physics, for all the empirical sciences, and of economics, for the social sciences as a whole), does not fit the variety of objects and problems that can be discussed, many of them completely new — in all the sciences, but particularly in the sciences of the artificial — that lead to new methods instead of a priori methods or the mere convergence of the methods already available.

Certainly, following a methodological pluralism, "scientists and philosophers should recognize that different descriptions and different approaches are sometimes beneficial because some descriptions offer better accounts of some aspects of a complex situation and other descriptions provide better accounts of other aspects" (Kellert et al. 2006b, xxiv). But to develop scientific activity requires going beyond description in order to get explanation or prediction (basic science), to obtain prediction to do prescription (applied science) or to be able to use the applied knowledge in the diverse contexts, according to the circumstances of each setting or case. Thus, the plurality of descriptions needs to be focused towards those scientific tasks.

(iv) Connected to the previous methodological consequences is the increasing difficulty of justifying a branch in any science that might be *fundamental* for the whole discipline, i.e., a part of a science that is the main or necessary support for the entire building of the discipline or offers the general scaffolding for the rest of the science and characterizes the key subject-matter for the research to be made. Thus, following pragmatic or pluralistic views in methodology of sciences makes it harder to propose any "fundamental" physics, chemistry, biology, psychology, etc.

(v) Other consequences — highlighted by methodological pragmatism — are the need to *assess* the efficacy of procedures and methods of scientific activity in engendering high quality results in the research made, rather than a kind of a priori configuration (monist, reductionist or universalist). This is in tune with the relevance of internal and external contextual factors for methodology of science[31] and the recognition of a plurality of stratagems for making research in empirical sciences (natural, social and of the artificial). Furthermore, there is the need for an ethical component while developing scientific methods, because values should have a role in scientific research.[32] This leads to multiscale modeling connected to levels of reality (micro, meso and macro).

1.3   Structure and Origin of this Book

In tune with the historical and thematic frameworks presented here of the pluralist and pragmatist methodological approaches, this book deals with the topic the methodological prospects for scientific research from different angles. It tries to offer a polyhedral perspective on science rather than a monist, reductionist or universalist conception of scientific activity. The main aim of this volume is to present a broader methodological scope, open to people with different backgrounds.

This can be seen in the structure of the volume, which has five parts: (I) A New Framework for Methodological Prospects to Scientific Research; (II) Pragmatist Approaches to Methodology of Science; (III) Contextual Factors for Methodology of Science; (IV) Methodological Pluralism in Natural Sciences and in Sciences of the Artificial; and (V) Methodological Pluralism in Social Sciences and Ethical Values. Each section has two papers. In this regard, parts I and II have a more general character than parts III, IV and V.

A new methodological framework begins with "Levels of Reality, Complexity, and Approaches to Scientific Method" by Wenceslao J. Gonzalez (University of A Coruña), which offers the *configuration of a new framework* for procedures and methods of scientific research. The *procedures* contribute to the initial stages of the scientific inquiry, whereas the *methods* enlarge our knowledge according to well-established ways or follow research processes that have been tested concerning their reliability. Scientific research needs methods that deal with objects and problems, whose diversity offers reasons for the unfeasibility of universal method for science and poses problems for methodological imperialism. The existence of levels of reality (micro, meso, and macro) and the features of structural and dynamic complexity pave the way for the methodological diversity. Thus, empirical sciences show different approaches to scientific method, according to the kind of science (natural, social and of the artificial) and also within sciences. Consequently, the

---

[31] The methodology of science deals with the advancement of knowledge, so, from an internal point of view, it presupposes the distinction between data, information and knowledge. From an external point of view, the methodology of science is the expression of the research activity of some agents in institutions (public or private).

[32] This also has consequences for technology, cf. Gonzalez (2015c).

relations of the scientific methods with the levels of reality and complexity require a deeper view than the conceptions already available.

"Multiscale Modeling: Explanation and Emergence" by Robert W. Batterman (University of Pittsburgh) offers a critique of methodological reductionism in physics, since in the physical phenomena studied in the text it is not necessary to start from a "Fundamental Physics." In this sense, there is a degree of autonomy in the scales of physical phenomena. It is — in his judgement — an explainable autonomy, so that this rethinks the question of emergence. On the basis of multiple realizability, there is room for success in research and autonomy of the continuum modeling. His conception leads to a theory that is "neither purely bottom-up nor purely top-down."

Then the pragmatist approach is available in "Methodological Pragmatism" by Nicholas Rescher (University of Pittsburgh), a key figure in this field. He focused his view on the efficacy of procedures and methods in engendering high quality products. This quality of the processes has to be assessed. In this regard, he considers that methodological pragmatism has the advantage of *greater reliability* due to the superior realism of the evaluations. I must stress that, in this text, Rescher expressly addresses the issue of the sphere of *applied science*, as opposed to the usual tendency in his conception to focus on basic science.

There is an analysis of Rescher's approach in "Methodological Incidence of the Realms of Reality: Prediction and Complexity" by Amanda Guillan (University of A Coruña). With a view from the ontology of science, the text insists that there are important limitations to the different varieties of methodological universalism. This paper takes into account the perspective of complexity and pays attention to the problems posed by methodological universalism to scientific prediction. On the one hand, the limits of methodological universalism are seen from the realms of reality, which leads to distinguish the natural, the social and the artificial. On the other, complexity — both structural and dynamic — requires attention, which leads to attention to modes of complexity and historicity.

After that a number of contextual factors for methodology of science are considered in "Information and Pluralism: Consequences for Scientific Representation and Method" by Giovanni Camardi (University of Catania). They belong to the "internal" context, insofar as the methodology of science the depends on epistemology, to the extent that it is based on information and representation. In this regard, a pluralist methodology comes from "inside," to the extent that it can be connected to a theory of information and seeks to interact with computational models. Based on the concept of information, the author considers possible the converge of Patrick Allo's notion of informational pluralism, Wesley Salmon's treatment of statistical relevance and James Woodward's analysis of the "data-phenomena" relationship.

Another angle in favor of pluralism based on contexts — this time from "outside" — is taken in "The Methodology of Theories in Context: The Case of Economic Clustering" by Catherine Greene and Max Steuer (London School of Economics). They argue that scientific theories are revealing in an "external" context (geographical, organizational, etc.) and, in the case of economic theories, they are incapable of universal application. One of the reasons is that they maintain that laws play little if any role in economics. In addition, they think that a single methodology of economics cannot be applicable to the variety of endeavours of economists. In the area of empirical inquiry, they see progress in the interaction between observations and theories, where "theories-in-context" is different from simple common sense. Economic clustering is used as an example of a pluralist approach to economic methodology.

Directly focused on pluralistic research strategies in nature is "Plurality of Explanatory Strategies in Biology: Mechanisms and Networks," by Alvaro Moreno (University of the Basque Country) and Javier Suárez (University of Barcelona). The focus here is connecting a pluralistic methodology of science and an ontology of science, i.e., the plurality of research strategies followed in life sciences and the existence of emergent levels of reality in that realm. Thus, the difference of explanatory strategies comes from the existence of an ontological diversity, which generates variations in causation. In this regard, mechanistic modelling and network modelling are connected to two ontological regimes of causation.

Another area where the pluralistic methodology of science meets an ontology is in "Scientific Prediction and Prescription in Plant Genetic improvement as an Applied Science of Design: The Natural and the Artificial" by Pedro Martínez Gómez (Higher Council of Scientific Research, Murcia). To exemplify methodological pluralism, the author turns to plant genetic improvement, which is methodologically dual: it is a science of nature, which analyses the genetic variation in the plant kingdom, and it is a science of the artificial, which is developed as an applied science of design to get new plants or different varieties. In addition, there are methodological differences in the levels involved: molecular biology, genetic constitution of the individuals and the release of new individuals. Prediction plays a key role, because it anticipates the possible future and serves the task of design. Again, internal and external factors related to the variables being studied need to be taken into account.

More specifically from a methodological point of view is the issue of validity, which is related to the link between the object of study, the problem posed and the solutions proposed. This consideration of methodology from "inside" leads to pluralism in the social sciences, as stated in "Challenges to Validity from the Standpoint of Methodological Pluralism: The Case of Survey Research in Economics" by María Caamaño-Alegre (University of Valladolid). She associates the development of a pluralistic methodology in economics with the possibility of improving the validity of empirical research. In this sense, the conditions of "internal" methodological validity are the preconditions for "external" methodological validity.

Any methodology of science does not deal with value-free research processes but with value-related processes, where ethical values have a place. This is particularly clear in methodology of economics, which is the axis of the paper "Economic Method and its Ethical Component: Pluralism, Objectivity and Values in Amartya Sen's Model" by Alessandra Cenci (University of Southern Denmark). This means that the plurality of economic methods requires an ethical component that overcomes the shortcomings of reductionist-monist economics.

Several of these papers were originally presented at the "Conference on Approaches to Scientific Method: Pluralism versus Reductionism" held at the University of A Coruña, Campus of Ferrol, on 12 and 13 March 2015. Thus, many of the topics are connected to the discussions held during those days. Like the XX Workshop on Contemporary Philosophy and Methodology of Science (*XX Jornadas sobre Filosofía y Metodología actual de la Ciencia*), it enlarges the topics already addressed in the set of volumes of previous workshops, which are gathered under the general title of *Gallaecia Series. Studies in Contemporary Philosophy and Methodology of Science.*[33]

It is my pleasure to thank the authors of the papers presented here for their efforts to offer new approaches to scientific method, taking into account the enlargement of the field in order to discuss novel topics. In addition to them and the other participants at the conference, I am very pleased to thank the editor of this *Synthese Series* — Otávio Bueno — for his interest in this new contribution to the ongoing discussions on the pragmatism and pluralism about scientific research.

Finally, my recognition again to the persons and institutions that cooperated in the original event. First, my appreciation to the Rector of the University of A Coruña, the Vice-rector of the Campus of Ferrol and Social Responsibility as well as to other academic authorities that have cooperated to this aim. Second, my acknowledgement to the organizations that gave their support: the Diputación of A Coruña, the City Hall of Ferrol, and the Santander Bank. My appreciation includes the Society of Logic, Methodology, and Philosophy of Science in Spain for its academic endorsement. In addition, let me express my gratitude to Jessica Rey, Pablo Vara and Amanda Guillan for their contribution to the edition of this book.

---

[33] This collection includes the following titles: *Progreso científico e innovación tecnológica* (1997), *El Pensamiento de L. Laudan. Relaciones entre Historia de la Ciencia y Filosofía de la Ciencia* (1998), *Ciencia y valores éticos* (1999), *Problemas filosóficos y metodológicos de la Economía en la Sociedad tecnológica actual* (2000), *La Filosofía de Imre Lakatos: Evaluación de sus propuestas* (2001), *Diversidad de la explicación científica* (2002), *Análisis de Thomas Kuhn: Las revoluciones científicas* (2004), *Karl Popper: Revisión de su legado* (2004), *Science, Technology and Society: A Philosophical Perspective* (2005), *Evolutionism: Present Approaches* (2008), *Evolucionismo: Darwin y enfoques actuales* (2009), *New Methodological Perspectives on Observation and Experimentation in Science* (2010), *Scientific Realism and Democratic Society: The Philosophy of Philip Kitcher* (2011), *Conceptual Revolutions: From Cognitive Science to Medicine* (2011), *Freedom and Determinism: Social Sciences and Natural Sciences* (2012), *Creativity, Innovation, and Complexity in Science* (2013), *Bas van Fraassen's Approach to Representation and Models in Science* (2014), *New Perspectives on Technology, Values, and Ethics: Theoretical and Practical* (2015), *The Limits of Science: An Analysis from "Barriers" to "Confines"* (2016), *Artificial Intelligence and Contemporary Society: The Role of Information* (2017), and *Philosophy of Psychology: Causality and Psychological Subject* (2018). See https:// cifcyt.udc.es/coleccion-gallaecia/ (Accessed on 9.5.2020).

References

Bergman, B. (2009). Conceptualistic pragmatism: A framework for Bayesian analysis? *IIE Transactions, 41*(1), 86–93.

Bertuglia, C. S., & Vaio, F. (2005). *Nonlinearity, chaos and complexity. The dynamics of natural and social systems*. Oxford: Oxford University Press.

Cartwright, N. (2004). Causation: One world, many things. *Philosophy of Science, 71*(5), 805–819.

Chang, H. (2012). *Is water H_2O? Evidence, realism and pluralism*. Dordrecht: Springer.

Dupré, J. (1993). *The disorder of things: Metaphysical foundations of the disunity of science*. Cambridge, MA: Harvard University Press.

Galavotti, M. C. (2008). Causal pluralism and context. In M. C. Galavotti, R. Scazzieri, & P. Suppes (Eds.), *Reasoning, rationality, and probability* (pp. 233–252). Stanford: CSLI Publications.

Galison, P., & Stump, D. J. (Eds.). (1996). *The disunity of science: Boundaries, contexts, and power*. Stanford: Stanford University Press.

Gelman, A. (2011). Bayesian statistical pragmatism. *Statistical Science, 26*(1), 10–11.

Giere, R. N. (2006a). *Scientific Perspectivism*. Chicago: The University of Chicago Press.

Giere, R. N. (2006b). Perspectival pluralism. In S. H. Kellert, H. E. Longino, & C. K. Waters (Eds.), *Scientific pluralism. XIX Minnesota studies in the philosophy of science* (pp. 26–41). Minneapolis: Minnesota University Press.

Gonzalez, W. J. (2006). Novelty and continuity in philosophy and methodology of science. In W. J. Gonzalez & J. Alcolea (Eds.), *Contemporary perspectives in philosophy and methodology of science* (pp. 1–28). A Coruña: Netbiblo.

Gonzalez, W. J. (2007). The role of experiments in the social sciences: The case of economics. In T. Kuipers (Ed.), *General philosophy of science: Focal issues* (pp. 275–301). Amsterdam: Elsevier.

Gonzalez, W. J. (2008). Rationality and prediction in the sciences of the artificial: Economics as a design science. In M. C. Galavotti, R. Scazzieri, & P. Suppes (Eds.), *Reasoning, rationality, and probability* (pp. 165–186). Stanford: CSLI Publications.

Gonzalez, W. J. (2011). From mathematics to social concern about science: Kitcher's philosophical approach. In W. J. Gonzalez (Ed.), *Scientific realism and democratic society: The philosophy of Philip Kitcher*, Poznan Studies in the Philosophy of the Sciences and the Humanities, (pp. 11–93). Amsterdam: Rodopi.

Gonzalez, W. J. (2012). Methodological universalism in science and its limits: Imperialism versus complexity. In K. Brzechczyn & K. Paprzycka (Eds.), *Thinking about provincialism in thinking*, Poznan Studies in the Philosophy of the Sciences and the Humanities (Vol. 100, pp. 155–175). Amsterdam/New York: Rodopi.

Gonzalez, W. J. (2013a). The roles of scientific creativity and technological innovation in the context of complexity of science. In W. J. Gonzalez (Ed.), *Creativity, innovation, and complexity in science* (pp. 11–40). A Coruña: Netbiblo.

Gonzalez, W. J. (2013b). From the sciences that philosophy has 'neglected' to the new challenges. In D. Dieks, W. J. Gonzalez, T. Uebel, M. Weber, & G. Wheeler (Eds.), *New challenges to philosophy of science* (pp. 1–6). Dordrecht: Springer.

Gonzalez, W. J. (2015a). *Philosophico-methodological analysis of prediction and its role in economics*. Dordrecht: Springer.

Gonzalez, W. J. (2015b). From the characterization of 'European philosophy of science' to the case of the philosophy of the social sciences. *International Studies in the Philosophy of Science, 29*(2), 167–188, available in: https://doi.org/10.1080/02698595.2015.1119418. Accessed on 25.1.2017.

Gonzalez, W. J. (2015c). On the role of values in the configuration of technology: From axiology to ethics. In W. J. Gonzalez (Ed.), *New perspectives on technology, values, and ethics: Theoretical and practical* (Boston Studies in the Philosophy and History of Science) (pp. 3–27). Dordrecht: Springer.

Gonzalez, W. J. (2017). From intelligence to rationality of minds and machines in contemporary society: The sciences of design and the role of information. *Minds and Machines, 27*(3), 397–424. https://doi.org/10.1007/s11023-017-9439-0.

Gonzalez, W. J. (2018). Complejidad dinámica en Internet como plataforma de información y comunicación: Análisis filosófico desde la perspectiva de Ciencias de Diseño y el papel de la predicción. *Informação e Sociedade: Estudos, 28*(1), 155–168.

Gonzalez, W. J. (2020a). Novelty in scientific realism: New Approaches to an Ongoing Debate. In: W. J. Gonzalez (Ed.), *New Approaches to Scientific Realism* (pp. 1–23). Boston/Berlin: De Gruyter. https://doi.org/10.1515/9783110664737-001.

Gonzalez, W. J. (2020b). Pragmatic realism and scientific prediction: The role of complexity. In: W. J. Gonzalez (Ed.), *New Approaches to Scientific Realism* (pp. 251–287). Boston/Berlin: De Gruyter. https://doi.org/10.1515/9783110664737-012.

Gonzalez, W. J. (2020c). Levels of reality, complexity, and approaches to scientific method. In W. J. Gonzalez (Ed.), *Methodological prospects for scientific research: From pragmatism to pluralism*, Synthese Library. Dordrecht: Springer.

Gonzalez, W. J., & Arrojo, M. J. (2019). Complexity in the sciences of the Internet and its relation to communication sciences. *Empedocles: European Journal for the Philosophy of Communication, 10*(1), 15–33.

Hacking, I. (1996). The disunities of the sciences. In P. Galison & D. J. Stump (Eds.), *The disunity of science: Boundaries, contexts, and power* (pp. 37–74). Stanford: Stanford University Press.

Humphreys, P. (2016). *Emergence: A philosophical account*. Oxford: Oxford University Press.

Kellert, S. H., Longino, H. E., & Waters, C. K. (Eds.). (2006a). *Scientific pluralism. XIX Minnesota Studies in the Philosophy of Science*. Minneapolis: Minnesota University Press.

Kellert, S. H., Longino, H. E., & Waters, C. K. (2006b). The pluralist stance. In S. H. Kellert, H. E. Longino, & C. K. Waters (Eds.), *Scientific pluralism. XIX Minnesota Studies in the Philosophy of Science* (pp. vii–xxix). Minneapolis: Minnesota University Press.

Kitcher, P. (2011a). Science in a democratic society. In W. J. Gonzalez (Ed.), *Scientific realism and democratic society: The philosophy of Philip Kitcher*, Poznan Studies in the Philosophy of the Sciences and the Humanities (pp. 95–112). Amsterdam: Rodopi.

Kitcher, P. (2011b). Scientific realism: The truth in pragmatism. In W. J. Gonzalez (Ed.), *Scientific realism and democratic society: The philosophy of Philip Kitcher*, Poznan Studies in the Philosophy of the Sciences and the Humanities (pp. 171–189). Amsterdam: Rodopi.

Kitcher, P. (2015). Pragmatism and Progress. *Transactions of C. S. Peirce Society, 51*(4), 475–494. https://doi.org/10.2979/trancharpeirsoc.51.4.06. (Accessed on 8.7.2019).

Longino, H. (1990). *Science as social knowledge*. Princeton: Princeton University Press.

Mainzer, K. (2007). *Thinking in complexity: The complex dynamics of matter, mind, and mankind* (5th ed.). New York: Springer.

Mitchell, S. (1992). On pluralism and competition in evolutionary explanations. *American Zoologist, 32*, 135–144.

Mitchell, S. (1993). *Biological complexity and integrative pluralism*. Cambridge: Cambridge University Press.

Mitchell, M. (2009). *Complexity: A guided tour*. Oxford: Oxford University Press.

Nagel, E. (1961). *The structure of science. Problems in the logic of scientific explanation*. New York: Harcourt, Brace and World.

Niiniluoto, I. (1993). The aim and structure of applied research. *Erkenntnis, 38*(1), 1–21.

Niiniluoto, I. (1999). *Critical Scientific Realism*. Oxford: Clarendon Press.

Niiniluoto, I. (2020). Interdisciplinarity from the perspective of critical scientific realism. In W. J. Gonzalez (Ed.), *New approaches to scientific realism* (pp. 231–250). Boston/Berlin: De Gruyter.

Popper, K. R. ([1956], 1992). Preface, 1956. In K. R. Popper, *Realism and the aim of science. From the postscript to the logic of scientific discovery: Vol. I*, edited by W. W. Bartley III, Hutchinson, London, 1983; reprinted by Routledge, London, 1992, pp. 5–8.

Psillos, S. (2009). Causal pluralism. In R. Vanderbreeken & B. D'Hooghe (Eds.), *Worldviews, science and us: Studies of analytical metaphysics. A selection of topics from a methodological perspective* (pp. 131–151). Singapore: World Scientific Publishers.

Rescher, N. (1973). *Conceptual Idealism.* Oxford: Blackwell (reprinted in Washington: University Press of America, 1982).

Rescher, N. (1977). *Methodological pragmatism: A system of pragmatic idealism.* Vol. I: *Human knowledge in idealistic perspective.* Oxford: Blackwell; New York: New York University Press.

Rescher, N. (1992). *A system of pragmatic idealism. Vol. I: Human knowledge in idealistic perspective..* Princeton: Princeton University Press.

Rescher, N. (1993). *Pluralism. Against the demand for consensus.* Oxford: Clarendon Press.

Rescher, N. (2012a). The problem of future knowledge. *Mind and Society, 11*(2), 149–163.

Rescher, N. (2012b). *Pragmatism: The restoration of its scientific roots.* New Brunswick: Transaction Publishers.

Rescher, N. (2014). *The pragmatic vision: Themes in philosophical pragmatism.* Lanham: Rowman and Littlefield.

Rescher, N. (2019). *Philosophical clarifications: Studies illustrating the methodology of philosophical elucidation.* Cham: Palgrave Macmillan.

Rescher, N. (2020). Methodological pragmatism. In W. J. Gonzalez (Ed.), *Methodological prospects to scientific research: From pragmatism to pluralism* (pp. 69–80), Synthese Library. Dordrecht: Springer.

Richardson, A. W. (2006). The many unities of science: Politics, Semantics, and Ontology. In S. H. Kellert, H. E. Longino, & C. K. Waters (Eds.), *Scientific pluralism. XIX Minnesota Studies in the Philosophy of Science* (pp. 1–25). Minneapolis: Minnesota University Press.

Rosser, J. B. Jr. ([1999], 2004). On the complexities of complex economic dynamics. *Journal of Economic Perspectives, 13*(4), 169–192. Reprinted in: J. B. Jr. Rosser (Ed.), *Complexity in Economics*, v. 1 (pp. 74–97). Cheltenham: Edward Elgar.

Simon, H. A. (1977). How complex are complex systems? In F. Suppe & P. D. Asquith (Eds.), *Proceedings of the 1976 biennial meeting of the Philosophy of Science Association* (Vol. 2, pp. 507–522). Ann Arbor: Edwards Brothers.

Simon, H. A. (1991). *Models of my Life.* N. York, NY: Basic Books.

Simon, H. A. (1996). *The sciences of the artificial* (3rd ed.). Cambridge, MA: The MIT Press, (1st ed., 1969; 2nd ed., 1981).

Simon, H. A. (1999). Can there be a science of complex systems? In Y. Bar-Yam (Ed.), *Unifying themes in complex systems: Proceedings from the international conference on complex systems 1997* (pp. 4–14). Cambridge, MA: Perseus Press.

Simon, H. A. (2001). Complex systems: The interplay of organizations and markets in contemporary society. *Computational and Mathematical Organization Theory, 7*(2), 79–85.

Suppe, F. ([1974] 1977). The search for philosophic understanding of scientific theories. In: F. Suppe (Ed.), *The structure of scientific theories* (pp. 1–241). Urbana: University of Illinois Press.

Suppes, P. ([1978] 1981). The plurality of science. In P. Asquith and I. Hacking (eds.), *PSA 1978*, East Lansing: Philosophy of Science Association, (vol. 2, 1981, pp. 3–16). (Reprinted in P. Suppes, *Probabilistic Metaphysics* (pp. 118–134). Oxford: B. Blackwell, 1984.)

Tiropanis, T., Hall, W., Crowcroft, J., Contractor, N., & Tassiulas, L. (2015). Network science, Web science, and Internet science. *Communications of ACM, 58*(8), 76–82.

Van Bouwel, J. (2004). Explanatory pluralism in economics: Against the mainstream. *Philosophical Explorations, 7*, 299–315.

Worrall, J. (2001). De la Matemática a la Ciencia: Continuidad y discontinuidad en el Pensamiento de Imre Lakatos. In W. J. Gonzalez (Ed.), *La Filosofía de Imre Lakatos: Evaluación de sus propuestas* (pp. 107–128). Madrid: UNED.

Part I
A New Framework for Methodological Prospects to Scientific Research

Chapter 2
Levels of Reality, Complexity, and Approaches to Scientific Method

Wenceslao J. Gonzalez

Abstract The methodological advances in science are above all associated with enhancing scientific knowledge by means of reliable processes. This requires the analysis of levels of reality, complexity and approaches to scientific method. In this regard, scientific processes can be procedures and methods. The procedures contribute to the initial stages of the inquiry and can complement the rigorous methods. Meanwhile, the methods enlarge our knowledge according to well-established ways or follow research processes whose reliability has been tested.

Scientific research needs methods that deal with objects and problems, whose diversity offers reasons for the unfeasibility of a universal method for science and poses problems for methodological imperialism. The existence of levels of reality (micro, meso, and macro) and the features of complexity, structural and dynamic complexity, pave the way for methodological diversity. Thus, empirical sciences show different approaches to scientific method, such as the differences between natural sciences and social sciences, and also the novelty of the sciences of the artificial in comparison with the social sciences. Consequently, the relations of the scientific methods with the levels of reality and complexity require a deeper view than the conceptions already available.

Keywords Levels of reality · Complexity · Approaches · Scientific method

This paper has been developed within the framework of the project FFI2016-79728-P, supported by the Spanish Ministry of Economics and Competitiveness (AEI).

W. J. Gonzalez (✉)
Center for Research in Philosophy of Science and Technology, Faculty of Humanities and Information Science, University of A Coruña, Campus of Ferrol, Ferrol, Spain
e-mail: wenceslao.gonzalez@udc.es

2.1   The Analysis of Approaches to Scientific Methods

Any approach to scientific method in empirical science deals with systematic ways of increasing scientific knowledge regarding reality — actual or possible — in order to get some kind of progress, which might be a *theoretical, empirical* or *heuristic* improvement.[1] Scientific method has more aspects involved than just the advancement of scientific knowledge (mainly, ontological and axiological facets, in addition to the epistemological factors), because science is a human activity geared towards specific goals and developed according to certain values. Nevertheless, the methodological growths in science are above all associated with enhancing scientific knowledge by means of reliable processes. Thus, they are effective — or even efficient — regarding some goals. These processes are related to a reality — natural, social or artificial — that has different levels and, very often, includes complexity.[2]

Accordingly, the analysis of the approaches to scientific methods can consider at least three kinds of aspects — epistemological, ontological and axiological — that can be related to the *assumptions, contents* or *limits of methods* in science. (a) There are some *assumptions* in scientific methods, which are connected to ontological issues (such as micro, meso or macro levels of reality or the intelligibility of phenomena[3]), to epistemological factors (such as cognitive rationality), and to axiological considerations (the kind of values — internal and external — preferred to choose ends and means of research). (b) The *contents* require cognitive rationality, while the *processes* themselves can be analyzed in connection with the practical rationality (or even with a possible logical basis shared by the processes) and the role of historicity.[4] The main values involved are efficacy and efficiency in problem-solving. The analysis of these processes can lead to a distinction between procedures and methods. (c) Any scientific method can have *limits* in two directions: in terms of a barrier, insofar as there should be some kind of frontier between a method that is scientific and another that is not, and in terms of confine, because there should

---

[1] On the present views on how to characterize "heuristic," see Chow (2015).

[2] This presupposes that there is a connection between methodology of science and ontology of science, which certainly has a close relationship with epistemology. In this regard, it is important to emphasize that the ontology of science is not something merely defended from perspectives of scientific realism but also from other philosophical positions, including anti-realist viewpoints. Thus, "even strongly empiricist approaches advocate a conception of scientific ontology: an ontology of observable objects, events, processes, and properties" (Chakravartty 2017, 41; see also 59–60 and 63).

[3] In this regard, it seems odd to claim that "the epithet 'intelligible' applies to theories, not to phenomena" (de Regt 2017, 12; see also pp. 45 and 88).

[4] Some of the conceptions in favor of monism, reductionism and methodological universalism, especially those of a directly logical-methodological kind, do not pay attention to historicity. But historicity is a key factor in understanding scientific change, complexity and problems related to scientific prediction, cf. Gonzalez, W. J. (2015b, 25, 29, 56n, 62, 77–78, 91, 103, 133, 185, 222–223, 249, 257, 267, 279n, 308 and 310).

be a ceiling for the method used (i.e., a realm where the method is actually valid) (Gonzalez 2016).

Altogether, this set of aspects configures the environment where the analysis of *levels of reality, complexity* and *approaches* to scientific method is located. In addition to these three features, there is a fourth one that accompanies them: the analysis needs to consider that there is a use of methods in accordance with the *kind of research*. Thus, there might be methodological differences between the research made in basic science, in applied science, and in the application of science.[5] Moreover, there might be distinctions between the methods in the cases of natural sciences, social sciences, and the sciences of the artificial.[6]

2.1.1 The Acceptance of Processes for Scientific Method: An Internal and an External Side

First of all, there is acceptance of *processes* in the approaches to scientific method, which includes an internal and an external side. From an *internal* viewpoint, these processes in science might be considered from different philosophical angles (universalist, pluralist, imperialist, convergent, selective,[7] etc.), which are assigned by the methodological approaches. But the processes that lead to scientific progress cannot be reduced to theoretical and empirical components, because heuristic aspects really matter if we want to grasp sciences such as psychology[8] or economics.[9]

From an *external* viewpoint, the approaches to scientific method can have an impact on the contexts where science has a relevant role, because science has a clear repercussion in contemporary society (in individuals, groups, organizations, nations, etc.), as the current research on Covid-19 patently illustrates. In addition, these approaches can have consequences for technological innovations (Gonzalez 2005), which can also lead to social innovation (as is the case with the sciences of the Internet and the electronic commerce) (Gonzalez 2020c).

Meanwhile the meta-methodological considerations take into account the existence of a *reality*, actual or possible, to be researched with the methods. In addition,

---

[5]Cognitive rationality with practical rationality and evaluative rationality are the three main spheres of rational deliberation, cf. Rescher (1988, 2–3). These three spheres of rationality may be intertwined in science. Moreover, this epistemic, practical and evaluative intertwining can be seen in the current efforts to find an effective treatment for Covid-19 patients and an adequate vaccine to avoid future problems. For pragmatism, they involve three ranges of philosophical concern, cf. Rescher (2019, 58).

[6]On the characterization of the sciences of the artificial, see Simon (1996).

[7]Convergent and selective prospects for methodology of science have been discussed in conceptions related to scientific realism. On these views, see Gonzalez (2020a).

[8]Regarding psychology, see for example Gigerenzer and Gaissmaier (2011).

[9]In the case of economics, this is particularly clear. See Gonzalez (2014).

the increase in scientific knowledge by means of reliable processes should have a clear support. Thus, either the processes are consistent in logical terms[10] or, at least, these processes are based on a neat rationality (i.e., a scientific rationality that includes historicity).[11] Commonly, the meta-methodological considerations assume that the whole set of elements involved with the scientific processes (theorizing, modeling, hypothesizing, representing, etc.) follow some kind of implicit or explicit rules, which are based on rational grounds (or even because they have some kind of logical basis).

2.1.2   A Framework of Processes: Procedures and Methods

These processes used in scientific research are at least twofold: procedures and methods. (a) "Procedures" are those processes that contribute to the initial stages of the inquiry (such as the processes for judgmental predictions) and can complement the rigorous methods (especially for qualitative factors). They might become genuine scientific methods in the future. Meanwhile, (b) the processes are already scientific "methods" when they enlarge our knowledge according to well established ways or follow research processes that have been tested for their reliability.[12]

But both kinds of processes used in science — procedures and methods — deal with an ontological basis — the levels of reality that configure the object studied — and they build upon the epistemological factors (in basic science, in applied science or in the application of science). Furthermore, they assume certain internal and external values, insofar as the ends and means of the processes of scientific research are chosen in accordance with certain values, which may be cognitive, social, etc.[13]

This difference regarding the *type of process* is particularly clear in the case of scientific prediction. We can distinguish two types of processes: (i) predictive procedures (i.e., informal or less sophisticated ways of searching), and (ii) methods of prediction (or "predictive methods" according to a more rigorous path). Both types of processes are used *de facto* in the daily practice of sciences such as economics. They can be employed independently or, as is more common, in a combined form. Moreover, the use of judgmental and scientific approaches is

---

[10] The search for a logical basis for methodological approaches is possible in the sense proposed by John Worrall. See Worrall (1988, 1989b, 1998).

[11] Scientific rationality should include historicity insofar as science is a human activity embedded by historicity in several ways. See Gonzalez (2011b, 2012c).

[12] On the difference between "procedures" and "methods" in science, which is noticeable in the case of scientific prediction, see Gonzalez (2015b, 255–273). On the setting of methodological options for scientific prediction, see Rescher (1998a, 85–112).

[13] These values play a role in preferring a type of methodological conception (pragmatic, pluralist, instrumentalist, etc.), but they also influence the configuration of the kind of social impact considered, which can be realistic, relativistic, constructivist, etc.

explicitly assumed in econometrics.[14] Indeed, in addition to the contribution of the economists, the role of non-specialists is sometimes accepted for adjustments.[15]

Among the reasons for using procedures and methods — and an explanation for the daily presence of both types of processes for prediction in science — is their need in solving problems related to complexity, mainly dynamic complexity. The advancement of science in any of these three cases of scientific research — basic, applied or of application — commonly requires dealing with complex systems (natural, social or artificial), which might be at least epistemological as well as ontological (Rescher 1998b, 9).

Moreover, this dual complexity can be structural and dynamic, which are two different aspects that are particularly relevant for scientific methods, as is seen in sciences such as biology (Gonzalez 2015c) and economics (Gonzalez 2011a, 2013b). Furthermore, the structural and the dynamic versions of complexity are open to a diversity of possibilities, according to the levels of reality (micro, meso and macro) and because of the variations of the temporal factors (in the case of prediction, there will be at least short, middle or long run).

2.2 Methods, Objects, and Problems

Initially, scientific methods are focused on two key factors: a *type of object*, either formal or empirical (i.e., natural, social or artificial), and a *sort of problem* — possible or actual — related to the object to be researched. In this regard, each approach to scientific method commonly requires one to take into account the level of reality studied (micro, meso, or macro) and within the temporal framework (past, present or future), which can include additional traits (such as short, middle or long run). In addition, the kind of complexity present in the research objects must very often be considered in problem-solving.

---

[14]"There are numerous ways of generating economic forecasts. Many are a mix of science — based on rigorously tested econometric systems — and judgment, occasioned by unexpected events: the future is not always like the present or the past" (Hendry and Ericsson 2001, 186).

[15]"Although progress is being made, we are still some way from a position where the model answers can be accepted without further human intervention. This is standard international practice. McNees surveyed the large U.S. forecasting organizations in 1981; they attributed between 20 and 50% of the final forecast to judgmental adjustments (. . .). Adjustments are made in the light of other information, commonsense judgements, past model error, and a knowledge of its deficiencies. The useful exercise of this judgement is not limited to the specialists. Non-specialists may also make a valuable contribution providing that the issues are put to them clearly" (Burns 1986, 104).

2.2.1   Complexity of the Objects to Be Studied and Perspectives for Research

Usually, the *complexity of the objects* to be researched can be studied in three ways: (a) in epistemological or ontological terms, (b) in its structural or dynamic varieties, and (c) its internal or external domains.[16] Following the type of object and the sort of problem, the research is led to a *scope* (i.e., a grade of generality in the problem to be studied, which varies from general to local) and which requires a *style of research* in tune with it. Thus, the research commonly takes the features of the accepted methods, or it proposes new methods to increase our knowledge, either of abstract forms (formal sciences) or of a possible or actual reality (empirical sciences).[17]

Consequently, a scientific method can be oriented to any of the three main *scientific ambits*: (1) basic science, where the focus is mainly theoretical, which leads to explanations or predictions; (2) applied science, where there is mostly a practical research, which includes predictions and prescriptions; and (3) application of science, which is that used by the agents of the scientific knowledge according to contexts (this contextual usage requires, in principle, the previous step of applied research but, at the same time, it can reinforce the research of applied science).[18]

Furthermore, the scientific methods can focus on the sort of problem to be researched from several possible *perspectives*. These may be commonly within five main cases: (i) that of "just" a discipline (such as physics, biology, economics, sociology, pharmacology, communication, etc.), (ii) that of an interdisciplinary field (such as biochemistry, psychophysiology, psychopedagogy, etc.),[19] (iii) that of a multidisciplinary endeavor (e.g., the studies of several sciences on the characteristics of a region or state), (iv) that of a crossdisciplinary task (such as conservation genetics),[20] or (v) that of a transdisciplinary province (such as gender studies).

---

[16]For complex systems, it is feasible to have a kind of methodological pluralism in terms of "having different models for different features of a phenomenon" (Morrison 2015, 7). But this involves the polyhedral character of the reality studied, whether natural, social or artificial.

[17]Obviously, "the abstract nature of mathematics can nevertheless yield concrete physical information" (Morrison 2015, 4).

[18]On the need for the distinction between applied science and application of science, see Niiniluoto (1993, 1–21; especially, 9 and 19). On the differences between basic science, applied science and application of science, see Gonzalez (2013a, 11–40; especially, 17–18).

[19]See monographic issue on "Philosophy of Interdisciplinarity," (2016). *European Journal for Philosophy of Science*, 6(3). It includes the paper by Knuuttila and Loettgers (2016). See also Niiniluoto (2020).

[20]Crossdisciplinarity is characterized by problems that are discussed using methodologies that, in principle, come from disciplines that are not thematically related. This is the case with a discipline at the micro level, such as genetics, and one at the macro level, such as environmental science, which intersect in the conservation genetics. Thus, thematic barriers are crossed in crossdisciplinarity, but not in principle methodologies. Meanwhile, methodologies are combined in interdisciplinarity, because a common point of encounter is sought from different starting points.

But the research object might be more complex than in previous decades, insofar as the intervention of technology is more intense, as in the case of the sciences of the artificial that deal with the development of the Internet as a network of networks (cf. Gonzalez 2018a),[21] where the communication and information technologies (ICT) have a crucial role in developing new aspects related to Artificial Intelligence (AI), such as "deep learning." This relation of scientific creativity and technological innovation can lead to new objects, novel problems, and scopes with no precedents, as happens with the research on AI and the development of computer sciences (Gonzalez 2017).

Another way to increase the complexity of the research object is *emergence*, which is related to the *novelty*, at least in two characteristic features. First, "there is some aspect of the emergent entity that is novel" (Humphreys 2016, 5), i.e., a characteristic that is autonomous regarding those components from which is developed. Second, there is some holism, insofar as there is at least one fact that, related to the new aspect, is not previously in what already exists and the rules that regulate it. Thus, emergence *adds something to the set* where the new appears (cf. Humphreys 2016, 5).

Methodologically, the emergence often has a transdisciplinary component, because emergence might be in a number of disciplines (physics, chemistry, biology, economics, sociology, etc.) that can be interrelated to give rise to something new within a certain set. Ontologically, emergence can be at any level of reality (micro, meso, macro) and can be found, in principle, in the natural, social and artificial realms. In the natural sphere, it commonly goes in two main directions: (a) transformational emergence (cf. Humphreys 2016, 60–69) and (b) fusion emergence (cf. Humphreys 2016, 70–86). In this regard, it is generally assumed that emergence occurs in an 'upward' direction, although Paul Humphries considers that "this is not inevitable" (Humphreys, 80).[22]

Besides the *level of reality* studied (micro, meso, or macro) and its complexity, scientific objects depend on the *temporal scale* of the phenomena (natural, social or artificial), where the short, middle or long run can be relevant. This is the case in *basic science* with explanations — especially if we work with retro-dictions — and in predictions (see, for example, Worrall 1989a). If we look backwards, then functional and genetic explanations combine the level of reality and the temporal scale in order to answer "why" questions (e.g., in biology or in demography). If we look forward, then foresights, predictions and forecasts require a specification

---

[21] This kind of science deals with designs and is different from the social sciences, even though there might be dual sciences — artificial and social — as happens with communication sciences, cf. Gonzalez (2008b).

[22] This downward or descendent approach seems odd, insofar as it suggests the ideas of immersion, submersion or submergence.

of the level of reality of the phenomena and temporal scale (which can include the immediate and the very long term).[23]

When the focus is on *applied science*, then levels of reality and temporal scales affect the relations between prediction and prescription, which are the main elements in modeling applied research (Simon ([1990] 1997). It seems clear that applied science requires solutions to concrete problems (Niiniluoto 1993, 1–21; especially, 2–3 and 5–6; and Niiniluoto 1995), where ontological factors are certainly important and are a condition for epistemological and methodological components. In addition, values have a role in order to choose what prescriptions should guide the solution of the concrete problems (as is seen in the case of economics) (Gonzalez 1998).

Also, *application of science* depends on the levels of reality. The ongoing debate on climate change has a very important component of the application of the scientific knowledge available, which includes a large set of predictions and a number of prescriptions to be implemented at the three levels: micro, meso, and macro. Thus, while some prescriptions can be implemented by individuals and small groups, others are to be carried out by organizations such as business firms and corporations, and there are also world-wide patterns of action that belong to the United Nations organization and the countries that signed the Paris agreement on 12 December 2015 (United Nations 2015).

2.2.2 *Unfeasibility of Universal Method for Science*

While doing research on this complex set of objects from quite different perspectives and on diverse levels of reality, we use a plurality of methods. That is why, although there might be a search for some central features shared by all the scientific methods used so far (and even those that might be used in an immediate future), there is no *universal method* that might be used in all the scientific areas.[24] Thus, formal and empirical sciences have differences regarding the type of problems, the kind of models (either descriptive or prescriptive)[25] and the characteristics of the results. In addition, the empirical sciences (natural, social and artificial) have methodological differences. This is also the case within the sciences themselves (physics, biology,

---

[23]Cf. Gonzalez (2015b, v, vii-viii, 2, 4, 6, 10n-11, 13, 18–21, 25, 30, 32–40, 47, 56, 60, 64, 69, 71, 77, 93, 114, 127, 129, 140, 150–151, 159, 165, 173, 184, 215–216, 218, 221, 249–250, 254–256, 264, 272, 275, 277–278, 288–289, 304, 308, 317–322, and 324–338).

[24]Cf. Gonzalez (2012a). The analysis made in this section and the next one is based on this paper.

[25]Descriptive models are characteristic of basic science, whereas prescriptive models are used in applied science.

economics, etc.), where there are methodological variations according to the level of reality studied (micro, meso and macro) and the problems addressed.[26]

Nonetheless, we can consider that there are varieties of methodological universalism. A key element is the extent of the scope, which allows several levels of methodological analysis. In principle, there are three *main ranks* of scientific research: (1) science in general (mainly, the empirical sciences)[27]; (2) a group of sciences, such as the natural sciences, the social sciences, or the sciences of the artificial; and (3) specific sciences, such as biology, economics, or computer sciences. These ranks embrace the diverse options of a methodological universalism in scientific research:

(a) Methodological universalism regarding *science* in general is now commonly considered problematic due, among other reasons, to the structural and dynamic complexity (Gonzalez 2020b). Thus, methodological proposals valid in principle for all empirical sciences, such as those of logical empiricism, are no longer accepted. (b) Another view is that of a methodological universalism in a *group of sciences*, as has happened in the social sciences, which has led to a version of methodological imperialism (based on economic imperialism). (c) There might be the view of a sort of methodological universalism located within the *specific sciences* (biology, economics, computer sciences, etc.).

Actually, we can see this version of universalism in disciplines of the natural sciences, such as biology (which is largely influenced by processes understood in evolutionary terms, frequently Darwinian); in subjects of the social sciences, such as economics (where the methods of mainstream economics — sometimes called "orthodox" economics — are still extremely influential both in theoretical and practical terms); and in studies of the sciences of the artificial, particularly in sciences of design developed in the sphere of computer sciences working with Artificial Intelligence.

What makes any sort of methodological universalism quite difficult is *complexity*. Certainly, organized complexity — and, even more, disorganized complexity — is a very important source of difficulties for methodological universalism. It can affect problems, methods and results of the scientific research. The features of complexity can be thought of as science, in general, a group of sciences, or a specific science, because complexity can be considered, in principle, by any of the disciplines related to nature, social and artificial worlds (see, in this regard, Mainzer 2007). To some extent, we can consider complex systems in these main spheres of the reality.[28]

---

[26]This feature of diversity is especially highlighted by methodological pluralism, but is also indicated by methodological pragmatism, when it connects the various goals sought with effectiveness in the research process. Cf. Gonzalez (2020e).

[27]Formal sciences, such as mathematics, commonly have specific methodological considerations, even though "quasi-empiricist" approaches and naturalist conceptions have searched for methodological similarities with empirical sciences.

[28]In addition, the *degree of complexity* matters, especially for modeling. Thus, in the case of computer simulation, "the system can be modelled at various levels of complexity, ranging from very simple models that don't include any interactions to more complicated modelling that

Complex systems can be focused either from the structural or the dynamic variety.[29] Both varieties involve the possibility of emergent properties.[30] In the first case, the study of complexity is commonly made regarding the framework or constitutive elements present in a group of sciences or in a specific science. Meanwhile, in the second — the dynamic version — addresses internal and external variations. In the dynamic case, the analysis of complexity is related to change over time of the motley elements involved in that collection of sciences or the specific science, taking into account the forces generating the change.[31]

2.2.2.1 Difficulties Based on Structural Complexity: Epistemological and Ontological

It seems clear that obstacles to methodological universalism can be located on both sides: in structural complexity and in dynamic complexity. Concerning the structural case, we consider the main epistemological and ontological aspects. Nicholas Rescher has made a relevant presentation (see, in this regard, Rescher 1998b, 9), where the *epistemic modes* of complexity are divided in three groups, in which it is possible to find a formulaic complexity: (i) descriptive complexity; (ii) generative complexity; and (iii) computational complexity.

Each presents epistemological difficulties for a universal method. De facto, they make it more difficult to characterize of the advancement of scientific knowledge that validly grasps how the finding and evaluation of science itself is, a group of sciences or a particular science. The difficulties can be in the three formulaic cases mentioned: (a) there might be complications in providing an adequate description of the complex system addressed; (b) there can be obstacles in providing the keys by which the system studied has been generated; and (c) there might be relevant issues of time and energy involved in solving the problems that such a complex system poses.

We can find *descriptive complexity* in each tier, if we accept that the configuration of science includes macro-theoretical frameworks, theories, models, hypotheses and processes for evaluation (observation, experimentation, ...). This difficulty of descriptive complexity increases when the object to be studied is macro and as broad as the biosphere, in the case of biology, or the infosphere,[32] in the case of the sciences of the artificial. Meanwhile, *generative complexity* can be at the

---

encompasses physics and engineering models, with the more complicated type giving rise to a greater probability errors" (Morrison 2015, 272–273).

[29] On complexity from a *dynamic point of view*, see Gonzalez (2013b).

[30] "The prospects for the emergence of an effective complex system are much greater if it has a nearly-decomposable architecture" (Simon 2001, 82).

[31] These categories of structural and dynamic can be used to articulate lists of kinds of complexity such as "multilevel organization, multicomponent causal interactions, plasticity in relation to context variation, and evolved contingency," Mitchell (2009, 21).

[32] On infosphere see Floridi (2014).

macro level (e.g., the cosmology around the big bang and the next steps), at the meso level (e.g., how certain epidemics are generated), and at the micro level (e.g., how social decline is generated in certain areas, according to the type of dominant industry). *Computational complexity* is characteristic of far-reaching and wide-ranging problems, whether spatial or temporal. This is the case for some astrophysical questions or issues related to climate change. But it can also occur when treating other problems, such as those related to the functioning of the human brain or issues regarding the human genome.

Even greater difficulties for the universal method come from the *ontological modes* of complexity. These are also distributed in three main groups: (1) compositional complexity; (2) structural complexity (in a strict sense); and (3) functional complexity. Within the compositional complexity, the possibilities are twofold: constitutional and taxonomical (or heterogeneity). Meanwhile, for Rescher, "structural complexity" also has two possibilities: it includes organizational and hierarchical complexities; whereas functional complexity is articulated into two modes: operational and nomic.

These ontological obstacles to a universal method can be in any type of reality (natural, social or artificial) and at any level (micro, meso or macro). Initially, there is the *complexity of composition*. One of the typical features of a complex system is that "the system cannot be fully understood by analysis into its components" (Humphreys 2016, 262). Thus, a biological ecosystem or an international social or economic organization is not reduced to its component parts. Constitutional complexity depends first on its components or constituent elements. Then there is a heterogeneity of elements, which expresses a taxonomical complexity in terms of the variety that can exist, for example, in a biological ecosystem or an international social or economic organization.

Thus, the *structural complexity* (in a strict sense) is also a source of obstacles. Organizational complexity can show many possible ways of arranging components, because they can be in different modes of interrelation (as the interdisciplinarity, transdisciplinarity or crossdisciplinarity make explicit). Even when there is hierarchical complexity, the possible relationship of subordination in terms of inclusion and subsumption can be questioned, as it happens nowadays with the criticism of the "fundamental" science. This is a problematic issue for some disciplines, which have traditionally worked with a part considered to be the bedrock of the discipline's architecture or that served as a general support for that branch of knowledge as a whole (physics, chemistry, biology, etc.).

Functional complexity can reveal varieties of modes of operation, which is one of the causes of concern in biomedicine and pharmacology. This is of particular concern for investigating viral pandemics such as Covid-19. In addition, the possible laws — if there are any in the phenomena studied — can be intricate. This is a particularly sensitive issue in the social sciences, where the question arises as to whether, first, "laws" fit in and, second, whether they are "laws of" or "laws in". When it comes to "laws of", it is considered that they are something that regulates the complex system under consideration (such as laws of the economy, like laws of the markets, etc., or laws of history, whether they are deterministic or not, etc.).

Whereas, if they are "laws in", then they are regularities or patterns found after a thorough analysis of a complex social system (such as the laws of economic transactions or the laws of demography).

2.2.2.2 Additional Difficulties: Dynamic Complexity Based on Historicity

In addition to structural complexity as a source of difficulties for methodological universalism, there is dynamic complexity, which brings with it the difficulties related to scientific change over time. Rescher's analysis is mainly related to structural complexity (the complex framework of the elements of science). He pays little attention to dynamic complexity (that connected to scientific change). Nevertheless, he is open to some dynamic aspects, which are relevant for science, in general, a group of sciences or a specific science. These dynamic aspects might be detected in the generative complexity (in the epistemic modes of complexity) and in the operational complexity as well as in nomic complexity (in the ontological modes of complexity).

Obviously, each mode of structural complexity — epistemic or ontological — can pose some difficulty for the universal methodology, regarding generality and reliability. But this difficulty increases when the dynamic complexity — due to changes through time — intervenes, which certainly modifies concepts — on the epistemological level — and which also generates ontological novelty (the emergence of a new property, of a different process, etc.). Throughout history of science this has sometimes translated into conceptual changes: the incorporation of new concepts, the change of sense of others already existing, ... and, when the variation is deep, we have conceptual revolutions (Thagard 1992).

Historically, it happens that the methodology of science needs to deal with issues that are not simple, which might be at different epistemological levels and can belong to diverse stages of reality. The researcher uses processes that depend on the objects (the aspect of reality studied) and the kind of problem (the focus of attention). In this regard, insofar as the scope of research is larger, the validity of the contents can, in principle, decrease due to the problems of testing the hypotheses. This can often be seen in economics, where dynamic complexity usually accompanies structural complexity (as evidenced by the previous international economic crisis and already seen with the current crisis due to Covid-19).

If we think of a science such as economics, the sources of *structural complexity* resemble a scale with several steps: (i) the social and artificial realms; (ii) the micro, meso, and macro levels; (iii) the degree of autonomy as human undertaking ("economic activity" and "economics as activity"); (iv) the organizations (big corporations, medium enterprises, small business firms, etc.) and markets (international,

national, regional, and local); (v) the role of groups and individual agents (i.e. creativity in different realms); ...[33]

Along with the multiscale structural complexity there is the *dynamic complexity*. When this happens (e.g., in economics), historicity has a key role and is another obstacle for methodological universalism. The change introduced by historicity — in knowledge and in reality — makes it more difficult to get universality across historical periods. This affects all kinds of objects of research: natural happenings, social events or contributions in the artificial world. Dynamic complexity also makes it difficult for there to be, for a long time, a dominant universalism in science itself, in a group of sciences or in a particular science. This is what happens if one accepts the existence of scientific revolutions, understood as profound changes in concepts, scientific practices and institutional approaches in research centers.

Through dynamic complexity there are then serious obstacles to two types of methodological universalisms. First, there are obvious obstacles for a methodological universalism conceived in a somehow "timeless" format, where the universality of the methodology is accepted across times. In this regard, dynamic complexity goes in the opposite direction to the conception that the scientific method can just be enlarged but not revised in any strong sense (e.g., the dominant view of logical positivism in the early stages of the Vienna Circle). Second, the dynamic complexity also makes it difficult for a temporal methodological universalism to last, when an approach is assumed as dominant during a historical period (e.g., what has happened in physics regarding methods used in mechanics in certain historical periods) (Gonzalez 2012a, pp. 159–161).

Certainly, a dynamic in terms of evolution (e.g., in the study of certain complex biological systems) is not the same as a dynamic conceived in terms of scientific *revolution* (e.g., in the transition from Newtonian to Einsteinian cosmology). But from the characteristics of historicity it is possible to encompass both "evolution" and "revolution" in science (Gonzalez 2011b). Also, as highlighted later, the historicity has a role in the *internal dynamics* in each of the steps of the scale indicated, so that it can be given in the realms, levels, types of activity, entities, agents (individuals or groups), ... In turn, historicity has a role in the *external dynamics*, since each of the steps of the scale has relations with the environment, whether natural, social or artificial.[34]

2.2.3 Problems for Methodological Imperialism

Another antagonist of the methodological diversity in science is the view of methodological imperialism. There is *methodological imperialism* in at least two

---

[33]The analysis of these elements is made in Gonzalez (2015b, especially chapter 7, 171–199), where there are more details about these issues.

[34]In the case of the sciences of the artificial, the dynamics is analyzed in Gonzalez (2013b).

different ways: intensity within a realm and extension to other realms. Thus, (I) methodological imperialism can be conceived of in terms of a kind of neat *prevalence* or clear *dominion* of some methods regarding a certain scientific realm (such as happened historically with Newtonian mechanics within physics as a whole); and (II) methodological imperialism can be understood as a set of methods that come from a *different discipline*, whose "boundaries" overflow to impinge on another field or fields (e.g., economic methods used in sociology, psychology, anthropology, law, political science, archeology, etc.). This predominance could be the case in any of these three levels of methodological analysis.

Again, there are several possibilities. Thus, methodological imperialism can be thought of as *science* in general, for example, developing a methodological proposal based on logical grounds and assuming the idea of universal validity of logic. But it is also possible to think of a methodological imperialism of a naturalist kind, which might be based, for example, on evolutionary grounds, such as the influential Darwinian approach (Gonzalez 2008c). Its repercussion is very noticeable in natural sciences, such as biology; in social sciences, such as in the evolutionary conceptions of psychology and economics (Nelson and Winter 1982; Hodgson 1993, 1995, 1999, 2001, 2004); and in the sciences of the artificial, such as computer sciences in terms of evolution of a complex system (Simon 1996, 188–190).

As a matter of fact, the attempt at a "methodological imperialism" has been made explicitly in the social sciences, while using economic methodology for solving very relevant social problems. This proposal for studying phenomena of a *group of sciences* has usually been connected to the work of a Nobel Prize winner in economics, Gary Becker,[35] a central figure of the Chicago school. Nonetheless, from a historical point of view, there are other authors that have been considered as supporters of an imperialism of economic roots.[36] Becker has tried to solve important social problems (e.g., those regarding family matters, such as marriages, divorces and fertility) by means of economic methods (based on neoclassical models).[37]

Economic imperialism, which is a form of economic expansionism, is a kind of methodological universalism that fits quite well into the influential tradition of "economic imperialism" defended at least by economists of the Chicago school. This view was accepted by George Stigler, who saw economics as an imperial science, insofar as "it has been aggressive in addressing central problems in a considerable number of neighboring social disciplines, and without any invitations" (Stigler 1984, 311).

---

[35]His view is analyzed in Pies and Leschke (1998). On methodological imperialism from a Popperian perspective, see Radnitzky and Bernholz (1987).

[36]This is the perspective that deals with domains of phenomena that previously were not generally perceived as "economic," but are now analyzed in economic terms. See Mäki (2009, 352).

[37]Among his most influential works are Becker (1976, 1981). On his views, see Cabrillo (1996). Regarding this topic, cf. Stigler (1984).

For Stigler, this noticeable influence of economics on other fields can be seen in four territories: law, history, social structure and behavior, and politics. Thus, methodological imperialism can be seen in at least in a number of cases: (1) the economics of law, with the application of economic analysis to legal rules and legal institutions, is in Ronald Coase and Richard Posner; (2) the new history made in economic terms is in Robert Fogel;[38] (3) the economics analysis of social structure and behavior (crime, racial discrimination, divorce, etc.) is developed by Gary Becker; and (4) the economic analysis of politics, for example of constitutional design, is used by James Buchanan and Gordon Tullock, the founders of the "Public Choice" school.[39] In all of them, the repercussion of economics is on a relevant scale and with a large number of specialists.

Undoubtedly, the problems posed by complexity in their several facets — epistemological and ontological, structural and dynamic, internal and external —[40] involve the existence of important methodological limits for this tradition of "economic imperialism." These have a direct repercussion on what methodological imperialism can actually achieve and lead to a methodological diversity in this group of sciences.[41] But this diversity is open to a possible convergence in the methodological components of scientific research. In this regard, the analysis of methodological diversity might show something that is shared by the *diversity of methods* used in science (natural, social, or the artificial).[42]

2.3 Levels of Reality and Complexity

One way to address methodological problems is in holistic terms, another is to deal with them according to levels of reality research. *Prima facie*, a methodological holism in science based on an ontological view without levels of reality has many problems, even in the case of physics. *De facto*, in the sphere of mechanics, physics distinguishes models that work well at the micro level of the atom (quantum mechanics), in the meso level of phenomena of movements on Earth (Newton's

---

[38]On Fogel — Nobel Prize winner in 1993 — and the methodology of the "new history," see Gonzalez (1996, 25–111; especially, 29, 37, 74–75, 86, 90–91, 95, 105, and 107).

[39]Buchanan was awarded the Nobel Prize in economics in 1986. Regarding his methodological views, and in particular his approach to prediction in economics, see Gonzalez (2006a, 89–90 and 100–101).

[40]On the internal and external complexity, see Gonzalez (2012b).

[41]The analysis of the relationship between economics and the Internet shows that it is a multivariate relationship. Thus, there are nuances in the role of economics depending on whether it is the scientific side, the technological facet or the social dimension of the network of networks. Cf. Gonzalez (2019).

[42]In the case of the sciences of the artificial related to the Internet, a common feature is the interdisciplinarity. See Tiropanis et al. (2015), Berners-Lee et al. (2006), and Hendler and Hall (2016).

fundamental laws) and in the macro level of the universe (Einstein's theory of relativity). Thus, a methodological approach needs to be aware of the role of the properties and processes that science grasps within our system of knowledge — properties and processes that are related to levels of reality.

2.3.1 Within the Levels of Reality: Properties and Processes

Methodological holism looks first at the studied whole and, within that whole, attends to its parts. Meanwhile, the levels of micro, meso and macro reality raise the possibility of differentiated properties and processes according to levels of reality (micro, meso and macro) and ontological realms (natural, social or artificial). In this respect, when investigating complex systems, we have the following:

(a) In many real world cases, "the whole is more than the sum of the parts,"[43] as Herbert Simon insisted for complex systems such as economics. This possibility is also valid for natural phenomena (e.g. biological organisms are more than the sum of their components) or even for complex artifacts (like in the case of the Internet, which with the Web and the cloud computing, works as a network of networks organized as a complex system with layers) (Clark 2018; Gonzalez and Arrojo 2019).

(b) There are "topological" properties, insofar as there are features of natural, social or artificial systems that appear only at a level of reality but are not available at other levels of the real world. In this regard, the laureates of the Nobel Prize in Physics 2016 have opened "the door on an unknown world where matter can assume strange states. They have used advanced mathematical methods to study unusual phases, or states, of matter, such as superconductors, superfluids or thin magnetic films. Thanks to their pioneering work, the hunt is now on for new and exotic phases of matter. Many people are hopeful of future applications in both materials science and electronics."[44]

(c) Complex systems are not commonly "isolated" structures but rather a type of structure interconnected with other structures of a different kind (as happens with

---

[43] Simon (1996, 184). In economics the complexity of the structures can originate emergent properties: Schenk (2006).

[44] The Royal Swedish Academy of Sciences, *The Nobel Prize in Physics 2016*, https://www. nobelprize.org/nobel_prizes/physics/laureates/2016/press.html (accessed on 1.12.2016). "The three Laureates' use of topological concepts in physics was decisive for their discoveries. Topology is a branch of mathematics that describes properties that only change step-wise. Using topology as a tool, they were able to astound the experts. (. . .) We now know of many topological phases, not only in thin layers and threads, but also in ordinary three-dimensional materials. Over the last decade, this area has boosted frontline research in condensed matter physics, not least because of the hope that topological materials could be used in new generations of electronics and superconductors, or in future quantum computers. Current research is revealing the secrets of matter in the exotic worlds discovered by this year's Nobel Laureates." The Royal Swedish Academy of Sciences (2016).

the physical properties connected to chemical properties, physical properties associated to biological properties or physiological elements related to psychological components). According to Humphries, "most sciences use a mixture of properties from other sciences" (Humphreys 2016, 126). What seems clear is that there might be properties of different kinds and that "multiscale modeling is common in areas such as climate modeling" (Humphreys 2016, 126).

(d) Complex systems can introduce new properties through changes over time. This kind of emergence may occur at the micro, meso or macro level, as happens in economics or in physics.[45] These emergent properties are due to internal and external variations over time. In this regard, it does not contribute much to the characterization of the variation introduced by the change over time to indicate that "it is an essential feature of emergence that the emergent entity emerge[s] from something else" (Humphreys 2016, 160).

Among the levels of reality, such as micro and macro, there might be interesting phenomena. Because "various kinds of materiality different systems such as magnets, fluids and gases having different kinds of micro-properties nevertheless exhibit the same kind of macro-behaviour when they reach their critical point" (Knuuttila and Loettgers 2016, 380). The existence of this kind of phenomena raises the issue of the relations between the levels of reality, which are three — micro, meso, and macro — instead of just two, because the properties at the low level might not be actually relevant for the high level (which happens with aggregate macroeconomic events in comparison with the microeconomic events) or even for the meso level. Meanwhile diverse phenomena at one of the levels can share some common properties even though they have different origins (which explains the use of evolutionism for the analysis of economic and sociological events).

2.3.2 Micro, Meso, and Macro as Ontological Levels for Complexity

Ontological levels of reality are relevant for methodological discussions. Thus, because of the interrelation between methods and objects of research, when science is dealing with methodological problems, the possible epistemological advancement depends on ontological support. This leads to different kinds of methods, insofar as a scientific method needs to be compatible with the level of reality it addresses. In this regard, ontological levels, such as micro, meso and macro, affect the set of methodological tasks of explanation and prediction (basic science), of prediction and prescription (applied science), and of the use of these methodological tools by the agents in the variety of contexts (application of science).

---

[45] Ontological emergence "asserts that genuinely novel objects and properties emerge even within the domain of physics, and it rejects the idea that only the level of fundamental physics is real" (Humphreys 2016, xvii).

2.3.2.1   The Distinction Focused on Rules

Sometimes the distinction between micro, meso and macro is used from an ontological viewpoint, where the focus is on rules (Dopfer et al. 2004) especially in the case of economics. Thus, "the central insight is that an economic system is a population of rules, a structure of rules, and a process of rules" (Dopfer et al. 2004, 263). Kurt Dopfer, John Foster and Jason Potts reduce meso to "a rule and its population of actualizations," meanwhile "micro refers to the individual carriers of rules and the system that they organize, and macro consists of the population structure of systems of meso" (Dopfer et al. 2004, 263).

However, this ontological architecture for a science, which is conceived as evolutionary (and, therefore, dynamic), seems to me to be partial, insofar as micro, meso and macro looks like a *purely methodological* approach that has ontological consequences. Moreover, in this view, the rules appear as more important than processes, which are certainly supposed by the rules. We can think of rules as part of organizations, which might be implicit or explicit, and are used in order to obtain some goals. Besides rules, there are genuine *ontological* components (individuals, groups, etc.), *epistemological* contents (organizations depend on knowledge to articulate the information available) and *values* (either implicit or explicit) that support the search of goals, the selection of means, and the evaluation of results.

A social ontology cannot be focused on a complete preeminence of rules, otherwise it is quite difficult to grasp many aspects, such as causality and causal explanation.[46] Primarily, the three ontological levels mentioned can be considered regarding basic science. Thus, if the scientific explanation is causal, then features of causality such as specificity, stability and proportionality depend on only one level of reality studied.[47] This is particularly clear in the features of specificity and proportionality in causality, because the relations between causes and effects require one to be on the same level of reality — micro, meso or macro — or, at least, to be on one level of reality that might be connected to the following one. The identification and individuation of causes in the micro level should be, in principle, easier to get than to do at the macro level, insofar as the degree of complexity is less intense.

2.3.2.2   The Case of Scientific Prediction

Scientific prediction does not depend only on rules and the temporal factor (such as the short, middle or long run) but also on the level of *reality* addressed by the predictive statements. Therefore, there are variations between prediction at the micro level, which might be particularly difficult (see Rescher 1998a, 1999), the meso level (such as multiple corporations and many medium size organizations), which has characteristics that might be more manageable than the prediction on

---

[46]On causality and causal explanation, see Gonzalez (2018b).

[47]These three features of causality appear in books such as Woodward (2003).

individuals or groups, and the macro level (such as nations or giant corporations and international organizations), where the size of the reality can make a better knowledge of the variables possible — in quantity and quality — than at the micro or meso levels.

Concerning *applied science*, the first step to guarantee is the scientific character of the predictions to be made. This means dealing with the *impediments* to predictability in any science. Commonly, they are studied in the case of basic science, but they might also concern applied science and they can spread from the application of science. These impediments are not mainly methodological but rather ontological and epistemological. *De facto*, the following main impediments are seen by Rescher in connection with *basic science*, and they are principally ontological and epistemological:

(1) Anarchy (i.e., lawlessness or absence of lawful regularities to serve as connecting mechanisms); (2) volatility (i.e., absence of nomic stability and of cognitively manageable laws); (3) uncertainty (i.e., the lack of information about the operative mechanisms); (4) haphazardness (i.e., the lawful linking mechanisms do not permit the secure inference of particular conclusions), which leads to three options: *chance and chaos*, *arbitrary choice*, or *change and innovation* (i.e., outcomes are not foreseeable because prediscernible patterns are continually broken); (5) fuzziness (i.e., data indetermination whether individually or in a collectively conjugate way); (6) myopia (i.e., data ignorance in the sense of lack of sufficient volume and detail to be able to make a prediction), and (7) inferential incapacity (i.e., the unfeasibility of carrying out the needed reasoning) (Rescher 1998a, 134–135).

Each impediment to predictability can be detected, in principle, at any level of reality (micro, meso, or macro). All of them have a relation with complexity — epistemological or ontological, either structural or dynamic, internal and external — because some of these impediments are mainly structural, whereas others are clearly dynamic. These impediments to predictability pose difficulties for a methodological universalism (Gonzalez 2012a) which is particularly relevant in biological sciences, where evolutionism has a dominant methodological role.[48]

Rescher recognizes that, for many writers, "complexity is determined by the extent to which chance, randomness, and lack of lawful regularity in general is absent" (Rescher 1998b, 8). But this concept of complexity, which is the inverse of simplicity, is an issue of *degree*: the system can be more or less complex. In the case of biological sciences, the tendency is to focus on some of the previous impediments to predictability, where uncertainty and haphazardness have, in principle, a relevant role from the methodological point of view.

---

[48] "'In ordinary English, a random event is one without order, predictability or pattern. The word connotes disaggregation, falling apart, formless anarchy, and fear.' This quote from the late Stephen J. Gould (1993) illustrates one reason why many nonbiologists — even highly educated ones — may feel uncomfortable with Darwinian evolution: Darwinian evolution centrally involves chance or randomness" (Wagner 2012, 95).

Obviously, complexity is the opposite to simplicity. But there are five possible concepts of simplicity: (i) parametric; (ii) theoretical; (iii) computational; (iv) epistemic; and (v) dimensional. These concepts are related to model selection (Rochefort-Maranda 2016, 261–279; especially, 269–274). Altogether these five versions of simplicity represent two main philosophical options of simplicity: epistemic and pragmatic (Rochefort-Maranda 2016, 261 and 274). The opposite notion — complexity — can be mainly epistemological and ontological with the structural and dynamic domains. But this complexity can be *pragmatic* in its models in two ways: (a) the scientific model is prepared with a purpose, i.e., it is oriented towards the resolution of a problem (theoretical, applied or of application); and (b) the model can be context-dependent and, therefore, connected to the agents.

Also, complexity in *applied science* is related not only to prediction but also to prescription, because any applied science requires giving the patterns for problem-solving (Gonzalez 2015b, ch. 12, 317–341). These patterns have a methodological component, but they also have epistemological and axiological components as can be noticed in the case of climate change (cf. Intermann 2015). Choosing the appropriate course of action for solving concrete problems (in physics, biology, economics, etc.) is not easy in many cases. Commonly, the complexity increases the difficulties of choosing the right course of action when we move from the micro level to the meso level or the macro level.

The next step is the *application of science*, which is clearly contextual and, therefore, pragmatic. When the issue of complexity is discussed in the sphere of the application of science, the levels of micro, meso and macro are particularly relevant. In addition, it is the issue of the kind of practical problem at stake (physical, economic, pharmacological, medical, etc.), which adds methodological diversity. This is particularly relevant in the case of medicine, where the relation between applied science (schools of medicine) and application of science (hospitals) is more intense and bi-directional. The micro level (a patient or a small group of patients), the meso (a larger group of persons with a common disease in diverse places) or macro (a pandemic with presence in several continents, as the case of Covid-19 emphasizes) are not the same.

2.3.3 Structural and Dynamic Complexity from the Internal-External Duality

Although the distinction between structural and dynamic complexity is key to this analysis of approaches to scientific method and the difficulties of methodological universalism, it does not cover the whole relevant field of complexity as it relates to the methodology of science. We also need the internal-external duality, which has epistemological content and ontological support. It is also connected to scientific values.

Complexity poses special difficulties for scientific methods due to the internal-external duality in scientific research, because the advancement of scientific knowledge for problem solving has to do with an environment with which it interacts. This makes it very difficult to have a methodological universalism "a priori," that is, a method of general validity prior to or outside of a connection with the complex reality being studied. In this regard, besides the differences in approaches between structural and dynamic dimension of complexity, which can be found in diverse realms of reality (natural, social, and artificial), there is the internal-external duality.

This duality of scientific activity for solving problems (basic, applied or application) is particularly relevant in the case of the dynamics. Thus, there is an *internal* facet of complexity in the complex systems, which is especially noticeable in the social and artificial systems (e.g., in the Internet as platform for human information and communication) (cf. Gonzalez 2018c), and an *external* trait of dynamic complexity related to complex systems, which may be in different spheres: economic, legal, social, political, etc. (Gonzalez and Arrojo 2015).

There is usually an internal and an external perspective in philosophy of science: the former has dominated for decades the philosophy of science, whereas the latter has had its leading role in the authors of the "social turn" and in the studies of science, technology and society (Gonzalez 2006b). Both philosophico-methodological lines — the internal and the external perspective — currently coincide in ruling out the possibility that there may be "the" scientific method in the singular and all-encompassing, one of the main reasons being the internal-external duality in the dynamics of complexity.

Following this duality of internal and external, the dynamic complexity is reinforced if we think of changes in complex systems as being due to *historicity*, instead of considering them in terms of certain generic processes or some kind of evolution, because historicity emphasizes the variability of complex systems, which can lead to a quite different state of affairs than at the beginning, as happens when there is a revolution (such as in the cases of social revolutions or the digital revolution). Moreover, complex systems can be organized (such as a hierarchical or poly-hierarchical system)[49] or disorganized (systems that might be chaotic, anarchic, etc.).

Historicity can be a central feature of the "internal" change of the complex system (particularly in social and artificial systems). In addition, it can also be a key trait of the "external" change of the complex system, because many complex systems are in a constant interaction with their environment (natural, social or artificial). This happens with the Internet, whose development — to a large extent — is due to the interaction with the users: individuals, groups, organizations, governments, … (Gonzalez 2020d). In addition, there are economic, legal, sociological, etc., aspects involved. Thus, a methodological approach needs to deal with internal and external dynamics of complex systems that are embedded with historicity.

---

[49]Simon is interested in hierarchical systems, whereas Stiglitz considers the poly-hierarchical systems as well. See Simon ([1973] 1977) and Sah and Stiglitz (1986).

2.4   Empirical Sciences and Approaches to Scientific Method

Contemporary methodological approaches cannot overcome the fact that there are three main groups of empirical sciences: natural, social, and artificial. All are relevant to complexity. For a long period (at least since the mid nineteenth century), the methodological comparison between natural sciences and social sciences has been frequent. Meanwhile, there has been little attention to the methodological comparison between social sciences and sciences of the artificial, in general, and the sciences of design, in particular.[50]

2.4.1   Differences Between Natural Sciences and Social Sciences

Methodologically, the differences between natural sciences and social sciences are mainly in the *Erklären-Verstehen* controversy, which has at least nine options (Gonzalez 2015a, 167–188; especially, 173–177). The differences between "explanation," conceived as the characteristic methodological approach to physics (and, thereafter, to the whole set of natural sciences), and "understanding," seen as the specific methodological approach to history (and, consequently, to the whole group of social sciences), have been discussed since 1858. In the succession of positions over the years, the nine options pointed out (J. G. Droysen-W. Dilthey, M. Weber, logical positivists, Wittgensteinians, H. G. Gadamer, G. H. von Wright, K. O. Apel, A. Giddens-Ch. Taylor-R. Bhaskar, and H. Lenk) include relevant details on the methodological approaches, including those related to procedures and methods.

Nevertheless, there are some aspects that, in one way or another, call the attention of those who wish to emphasize that natural phenomena and social events require different methodological approaches. These discrepancies directly affect the methods of basic science and, thereafter, the methods of applied science but are also relevant to the application of science. *De facto*, although many philosophers and scientists accept a common methodological ground between both kinds of sciences, the *Erklären-Verstehen* controversy emphasizes the several methodological differences between them, which are connected to epistemological and ontological issues:

(a) Natural methods are commonly focused on "universal" phenomena in cases such as physical laws, which can usually be repeated, whereas social methods often need to deal with singular events in history, which in principle cannot be repeated in a strict sense (such as a battle or the death of a key leader). (b) Natural phenomena are usually beyond our scope in terms of intervention ('we cannot decree a rainfall' or 'we cannot stop a hurricane'), whereas social events are led by intentionality and

---

[50]On the sciences of the artificial from the perspective of the sciences of design, the most influential book is Simon's volume mentioned already, whose third edition was published in 1996. An analysis of the case of economics is in Gonzalez (2008a).

normativity,[51] insofar as social events are related to human actions — individual and social — and they are oriented toward ends through decision-making and within a setting of implicit or explicit rules.

(c) The possibility of achieving objectivity by means of natural methods seems to be ordinarily guaranteed, especially in physics, whereas the acceptance of something as objective as result of the research in social sciences is often controversial or even explicitly denied, mainly in cases such as historical events. (d) The status of the relations between researcher and the reality researched in the case of natural phenomena seems quite different from those in the case of social events, insofar as in social affairs the subject who does the research of the event is or might be at the same time — at least, to some extent — part of the object that is researched.

(e) Although natural phenomena require a context of interpretation of the data available and informative statements (e.g., in earthquakes we need a scale of interpretation of data), the role of interpretation is commonly stronger in the social sciences than in natural sciences. In this regard, some methodological views claim that the social events are usually built on a first kind of interpretation in order to characterize a social fact (and, therefore, it might be a plain construction),[52] followed by a second kind of interpretation that inserts the social event into a wider framework (social, cultural, economic, cultural, etc.).

Given these methodological differences between natural sciences and social sciences, we do not exhaust the whole set of possible differences. Additional features can be found if we consider basic science, applied science, and application of science. In the area of *basic science*, the differences can be extended to the characterization of the explanations when they are causal[53] or to the obstacles for prediction, which increase in social events in comparison with natural phenomena.

If we look at the sphere of *applied science*, there are also differences between natural sciences and social sciences in the relation between prediction and prescription, which are particularly clear in the case of prescription of patterns for solving problems due to social events, because scientific prescriptions (for example, in economics or demography) require special assessments based on the acceptance of values.[54]

---

[51] The role of normativity, see Spohn (2011).

[52] On this issue, see Hacking (1999).

[53] The ongoing discussions on causality (such as actual causation, causal selection — one or several causes — and causal importance) might lead to differences between the cases of natural phenomena and social events. In addition, the features of causal importance might be of a different kind in natural phenomena than in social events. This concerns several aspects: (i) the causal responsibility (what produces the effect and makes the trait in the effect possible), (ii) the difference in the making of the effect, either actual or potential, and (iii) the causal specificity. Thus, natural intervention in physical phenomena and the agent intervention in economic events can have different characteristics.

[54] "Prediction is not the only exercise with which economics is concerned. Prescription has always been one of the major activities in economics, and it is natural that this should have been the case. Even the origin of the subject of political economy, of which economics is the modern version, was clearly related to the need for advice on what is to be done in economic matters. Any prescriptive

In addition, there are also methodological differences in the *application of science*, because in the case of social sciences the processes of application of knowledge need ethical values (e.g., in psychology or psychiatry).[55] This shows the relevance of considering pragmatic complexity in addition to structural and dynamic complexity.

2.4.2 The Novelty of the Sciences of the Artificial in Comparison with the Social Sciences

Overall, there is a constitutive feature of novelty in the sciences of the artificial, insofar as they look for an extension or enlargement of human possibilities according to new aims, novel processes, and innovative results. Meanwhile the social sciences are commonly related to human needs, which seek universal features (social, cultural, political, etc.) in specific spheres. Thus, although the social sciences include many expressions of variation, they share some type of common roots that can be found in the past and the present, and they will appear again in the future in diverse forms.

Novelty in the sciences of the artificial, which are commonly applied sciences and have many applications, can appear in two main directions: horizontal and vertical. It is horizontal when the expansion is based on something previously available that is enlarged in order to reach new aims. It is vertical when the research moves up to get new aims that were not possible in the past, either as a new combination of existing elements (creativity as a reassemble of elements where new properties emerge) or as full-fledged creation of new elements[56] (as happens many times with the creativity connected with the Internet).[57]

Horizontal (or longitudinal) and vertical (or transversal) ways of novelty in the realm of the artificial can appear at the micro, meso or macro level. They can be seen in disciplines based on designs, such as communication sciences, economics or information science.[58] Within the micro level, a horizontal way of novelty is in the designs that enable faster communication between individuals or small groups, to get new forms of microcredits in economics or to achieve innovative paths

---

activity must, of course, go well beyond pure prediction, because no prescription can be made without evaluation and an assessment of the good and the bad," Sen (1986, 3).

[55] In the sphere of medicine, there is a clear connection between methodological problems and ethical values, which affect applied science and the application of science. In this regard, see Worrall (2006).

[56] A very innovative author in the sphere of the sciences of the artificial was Alan Turing, a key figure in the development of the Artificial Intelligence. See Hodges (2014).

[57] See, for example, the case of the Web Science, cf. Berners-Lee et al. (2006), Hendler and Hall (2016), and Tiropanis et al. (2015).

[58] All of them share designs that look for aims, being followed by processes in order to reach expected results. At the same time, they deal with complex systems. See Gonzalez (2011a, 2012b) and Gonzalez and Arrojo (2019).

to organize some libraries with books. Meanwhile, a vertical way of novelty at the micro level is in the programs for payments by cell phone instead of using checks or bank notes, the use of e-mails as a substitute for traditional letters, or the organization of the personal digital library of e-books and e-journals.

If the level is meso, we can find horizontal novelty in the programing for digital radio, in the design of marketing for a regional or state digital television, and in the scientific resources of the cell phone for direct intercontinental calls. Vertical novelty at the meso level appears in the programs for the Intranet of business firms or corporations, in the design of web pages as institutional images of a public or private university or center of research, or the organization of an electronic network of libraries using the same cataloguing system.

An additional degree of complexity comes from the horizontal novelty when the level is macro, such as the programing of digital television that has world-wide coverage (like television over the top, OTT), the design of financial products by the Federal Reserve or the European Central Bank that takes into account economic history (such as the 1929 crisis) to deal with the recent crisis, which began around 2007, and the designs for a European network of public libraries. Vertical novelty is in the new forms of communication based on the Internet, such as the design of social networks (either oriented towards friends — Facebook, Snapchat, etc. — or with a professional approach — LinkedIn, Mendeley, etc. —) or new informative means (Twitter, YouTube, etc., and those developed in China and India). Furthermore, all the expressions of financial economics related to international e-commerce include vertical novelty, and this kind of novelty is also in the designs of repertoires of bibliographical information with millions of references (such as Scopus).

Besides the twofold horizontal-vertical novelty, there is another duality that is methodologically relevant: internal novelty and external novelty. The novelty is "internal" or endogenous when the variations are made because of new processes needed for reaching new aims in order to get novel results (e.g., in the field of Artificial Intelligence or in the sciences of the Internet). The novelty is "external" or exogenous when the changes come from several sources: (a) the demands of the users (in economics through new accessibility for individuals, business firms, or international organizations for economic transactions; in communication through new programs for information, entertainment, etc.); (b) some legal dispositions (such as right to privacy or the "right to be forgotten," etc.); (c) the new possibilities introduced by novel technological devices; etc.

All these forms of novelty due to the sciences of the artificial include aspects that are related to something that is *de facto* optional instead of being really needed. Thus, they add something that it is not merely social in the sense of based on human needs. This connects with what Ortega y Gasset used to call "supernature" (*sobrenaturaleza*), i.e., something that is added to what is really natural and, therefore, is above what is originally natural.[59] Thus, artificial is designed in order to grasp new possibilities for human beings, which we can see every day with the use of the Internet. This realm of the artificial is based on synthesis, whereas the social realm is developed with the role of analysis.[60]

2.5 Coda

According to the analysis made here, the relations of the scientific methods with the levels of reality and complexity require a deeper view than the conceptions already available. (i) Methodological approaches are intertwined with the kinds of components: *ontological* (an object within a level of reality), *epistemological* (the advancement of knowledge in a basic, applied or of application setting), and *axiological* (the values that lead to preferences on the selection of ends and means of research). (ii) There is a *methodological diversity*, which starts with differences between the methods in formal sciences and empirical sciences, followed by differences in the methods used in natural, social, and artificial sciences. Thus, there is no basis for a methodological uniformity, either universal or sectorial, in science. In addition, methodological imperialism is problematic, even though there is an increasing need for interdisciplinary, multidisciplinary, crossdisciplinary and transdisciplinary studies in science.

   (iii) Methodological diversity and *ontological levels* (micro, meso, and macro) change the dominant view of an architectonic of science in terms of a building, where there are some fundamental parts understood as the basis for the bricks of knowledge of the other parts of the discipline. Thus, there is no genuine "fundamental" physics, chemistry, biology, etc., followed by the other disciplines in such an area of research. In addition, this new view fits in with the existence of phenomena such as the four forces of physics without a unified force to put them altogether in a consistent way. (iv) Complexity highlights the need for a *pragmatic* component of adaptation of methods to objects, where the objects are at a level of reality (micro, meso, or macro) and in a context of problems interrelated with other problems. This pragmatic realist view is in tune with a structural complexity and a dynamic complexity, where the models used — descriptive or prescriptive — can vary from one domain, such as natural sciences, to another domain, either social sciences or the sciences of the artificial.[61]

---

[59] José Ortega y Gasset used this concept for the philosophy of technology, Ortega y Gasset ([1933] 1997, 23, 24 and 60). To some extent, it is a feature that is also valid for artificial designs, insofar as they enlarge the possibilities of what is natural in the human agents and societies in order to get new aims.

[60] Simon insisted on the feature of synthesis for the sciences of the artificial. See Simon (1996, 4–5).

[61] A development of this approach within the framework of scientific realism can be found in Gonzalez (2020b).

References

Becker, G. S. (1976). *The economic approach to human behavior*. Chicago: The University of Chicago Press.

Becker, G. S. (1981). *A treatise on the family*. Cambridge, MA: Harvard University Press.

Berners-Lee, T., Hall, W., Hendler, J., Shadbot, N., & Weitzner, D. J. (2006). Creating a science of the Web. *Science, 313*(5788), 769–771.

Burns, T. (1986). The interpretation and use of economic predictions. In J. Mason, P. Mathias, & J. H. Westcott (Eds.), *Predictability in science and society* (pp. 103–125). London: The Royal Society and The British Academy.

Cabrillo, F. (1996). *The economics of family and family policy*. Cheltenham: E. Elgar.

Chakravartty, A. (2017). *Scientific ontology: Integrating naturalized metaphysics and voluntarist epistemology*. New York: Oxford University Press.

Chow, S. J. (2015). Many meanings of 'Heuristic'. *British Journal for Philosophy of Science, 66*(4), 977–1016.

Clark, D. D. (2018). *Designing an Internet*. Cambridge, MA: The MIT Press.

de Regt, H. W. (2017). *Understanding scientific understanding*. Oxford: Oxford University Press.

Dopfer, K., Foster, J., & Potts, J. (2004). Micro-meso-macro. *Journal of Evolutionary Economics, 14*(3), 263–279.

Floridi, L. (2014). *The fourth revolution – How the Infosphere is reshaping human reality*. Oxford: Oxford University Press.

Gigerenzer, G., & Gaissmaier, W. (2011). Heuristic decision making. *Annual Review of Psychology, 62*, 451–482.

Gonzalez, W. J. (1996). Caracterización del objeto de la Ciencia de la Historia y bases de su configuración metodológica. In W. J. Gonzalez (Ed.), *Acción e Historia. El objeto de la Historia y la Teoría de la Acción* (pp. 25–111). A Coruña: Publicaciones Universidad de A Coruña.

Gonzalez, W. J. (1998). Prediction and prescription in economics: A philosophical and methodological approach. *Theoria, 13*(32), 321–345.

Gonzalez, W. J. (2005). The philosophical approach to science, technology and society. In W. J. Gonzalez (Ed.), *Science, technology and society: A philosophical perspective* (pp. 3–49). A Coruña: Netbiblo.

Gonzalez, W. J. (2006a). Prediction as scientific test of economics. In W. J. Gonzalez & J. Alcolea (Eds.), *Contemporary perspectives in philosophy and methodology of science* (pp. 83–112). A Coruña: Netbiblo.

Gonzalez, W. J. (2006b). Novelty and continuity in philosophy and methodology of science. In W. J. Gonzalez & J. Alcolea (Eds.), *Contemporary perspectives in philosophy and methodology of science* (pp. 1–28). A Coruña: Netbiblo.

Gonzalez, W. J. (2008a). Rationality and prediction in the sciences of the artificial: Economics as a design science. In M. C. Galavotti, R. Scazzieri, & P. Suppes (Eds.), *Reasoning, rationality, and probability* (pp. 165–186). Stanford: CSLI Publications.

Gonzalez, W. J. (2008b). La televisión interactiva y las Ciencias de lo Artificial. In M. J. Arrojo (Ed.), *La configuración de la televisión interactiva: De las plataformas digitales a la TDT* (pp. xi–xvii). A Coruña: Netbiblo.

Gonzalez, W. J. (2008c). Evolutionism from a contemporary viewpoint: The philosophical-methodological approach. In W. J. Gonzalez (Ed.), *Evolutionism: Present approaches* (pp. 3–59). A Coruña: Netbiblo.

Gonzalez, W. J. (2011a). Complexity in economics and prediction: The role of parsimonious factors. In D. Dieks, W. J. Gonzalez, S. Hartman, T. Uebel, & M. Weber (Eds.), *Explanation, prediction, and confirmation* (pp. 319–330). Dordrecht: Springer.

Gonzalez, W. J. (2011b). Conceptual changes and scientific diversity: The role of historicity. In W. J. Gonzalez (Ed.), *Conceptual revolutions: From cognitive science to medicine* (pp. 39–62). A Coruña: Netbiblo.

Gonzalez, W. J. (2012a). Methodological universalism in science and its limits: Imperialism versus complexity. In K. Brzechczyn & K. Paprzycka (Eds.), *Thinking about provincialism in thinking* (Poznan Studies in the Philosophy of the Sciences and the Humanities) (Vol. 100, pp. 155–175). Amsterdam/New York: Rodopi.

Gonzalez, W. J. (2012b). Las Ciencias de Diseño en cuanto Ciencias de la Complejidad: Análisis de la Economía, Documentación y Comunicación. In W. J. Gonzalez (Ed.), *Las Ciencias de la Complejidad: Vertiente dinámica de las Ciencias de Diseño y sobriedad de factores* (pp. 7–30). A Coruña: Netbiblo.

Gonzalez, W. J. (2012c). La vertiente dinámica de las Ciencias de la Complejidad. Repercusión de la historicidad para la predicción científica en las Ciencias de Diseño. In W. J. Gonzalez (Ed.), *Las Ciencias de la Complejidad: Vertiente dinámica de las Ciencias de Diseño y sobriedad de factores* (pp. 73–106). A Coruña: Netbiblo.

Gonzalez, W. J. (2013a). The roles of scientific creativity and technological innovation in the context of complexity of science. In W. J. Gonzalez (Ed.), *Creativity, innovation, and complexity in science* (pp. 11–40). A Coruña: Netbiblo.

Gonzalez, W. J. (2013b). The sciences of design as sciences of complexity: The dynamic trait. In H. Andersen, D. Dieks, W. J. Gonzalez, T. Uebel, & G. Wheeler (Eds.), *New challenges to philosophy of science* (pp. 299–311). Dordrecht: Springer.

Gonzalez, W. J. (2014). The evolution of Lakatos's repercussion on the methodology of economics. *HOPOS: The Journal of the International Society for the History of Philosophy of Science, 4*(1), 1–25.

Gonzalez, W. J. (2015a). From the characterization of 'European Philosophy of Science' to the case of the philosophy of the social sciences. *International Studies in the Philosophy of Science, 29*(2), 167–188.

Gonzalez, W. J. (2015b). *Philosophico-methodological analysis of prediction and its role in economics*. Dordrecht: Springer.

Gonzalez, W. J. (2015c). Prediction and prescription in biological systems: The role of technology for measurement and transformation. In M. Bertolaso (Ed.), *The future of scientific practice: 'Bio-Techno-Logos'* (pp. 133–146 text and 209-213 notes). London: Pickering and Chatto.

Gonzalez, W. J. (2016). Rethinking the limits of science: From the difficulties to the frontiers to the concern about the confines. In W. J. Gonzalez (Ed.), *The Limits of Science: An Analysis from "Barriers" to "Confines"* (Poznan Studies in the Philosophy of the Sciences and the Humanities) (pp. 3–30). Leiden: Brill-Rodopi.

Gonzalez, W. J. (2017). From intelligence to rationality of minds and machines in contemporary society: The sciences of design and the role of information. *Minds and Machines, 27*(3), 397–424. https://doi.org/10.1007/s11023-017-9439-0.

Gonzalez, W. J. (2018a). Internet en su vertiente científica: Predicción y prescripción ante la complejidad. *Art, 7*(2). 2nd period, 75–97. https://doi.org/10.14201/art2018717597.

Gonzalez, W. J. (2018b). Configuration of causality and philosophy of psychology: An analysis of causality as intervention and its repercussion for psychology. In W. J. Gonzalez (Ed.), *Philosophy of psychology: Causality and psychological subject. New reflections on James Woodward's contribution* (pp. 21–70). Boston/Berlin: Walter de Gruyter.

Gonzalez, W. J. (2018c). Complejidad dinámica en Internet como plataforma de información y comunicación: Análisis filosófico desde la perspectiva de Ciencias de Diseño y el papel de la predicción. *Informação e Sociedade: Estudos, 28*(1), 155–168.

Gonzalez, W. J. (2019). Internet y Economía: Análisis de una relación multivariada en el contexto de la complejidad. *Energeia: Revista internacional de Filosofía y Epistemología de las Ciencias Económicas, 6*(6), 11–36. Available at: https://abfcfc9a-c7ef-4730-b66e-0a415ef434c0.filesusr.com/ugd/e46a96_b400af5a739e4310a31b7e952244745d.pdf. Accessed on 1.4.2020.

Gonzalez, W. J. (2020a). Novelty in scientific realism: New approaches to an ongoing debate. In W. J. Gonzalez (Ed.), *New approaches to scientific realism* (pp. 1–23). Boston/Berlin: De Gruyter. https://doi.org/10.1515/9783110664737-001.

Gonzalez, W. J. (2020b). Pragmatic realism and scientific prediction: The role of complexity. In W. J. Gonzalez (Ed.), *New approaches to scientific realism* (pp. 251–287). Boston/Berlin: De Gruyter. https://doi.org/10.1515/9783110664737-012.

Gonzalez, W. J. (2020c). Electronic economy, internet and business legitimacy. In J. D. Rendtorff (Ed.), *Handbook of business legitimacy: Responsibility, ethics and society*. Dordrecht: Springer.

Gonzalez, W. J. (2020d). La dimensión social de Internet: Análisis filosófico-metodológico desde la complejidad. *Artefactos: Revista de Estudios sobre Ciencia y Tecnología, 9*(1), 101–129. https://doi.org/10.14201/art2020101129.

Gonzalez, W. J. (2020e). Pragmatism and pluralism as methodological alternatives to monism, reductionism and universalism. In W. J. Gonzalez (Ed.), *Methodological prospects for scientific research: From pragmatism to pluralism*, Synthese Library (pp. 1–18). Dordrecht: Springer.

Gonzalez, W. J., & Arrojo, M. J. (2015). Diversity in complexity in communication sciences: Epistemological and ontological analyses. In D. Generali (Ed.), *Le radici della razionalità critica: Saperi, Pratiche, Teleologie* (Vol. I, pp. 297–312). Milan-Udine: Mimesis.

Gonzalez, W. J., & Arrojo, M. J. (2019). Complexity in the sciences of the Internet and its relation to communication sciences. *Empedocles: European Journal for the Philosophy of Communication, 10*(1), 15–33. https://doi.org/10.1386/ejpc.10.1.15_1.

Hacking, I. (1999). *The social construction of* what? Cambridge, MA: Harvard University Press.

Hendler, J., & Hall, W. (2016). Science of the world wide web. *Science, 354*(6313), 703–704.

Hendry, D. F., & Ericsson, N. R. (2001). Epilogue. In D. F. Hendry & N. R. Ericsson (Eds.), *Understanding economic forecasts* (pp. 185–191). Cambridge, MA: The MIT Press.

Hodges, A. (2014). *Alan Turing: The enigma*. London: Vintage Books/Random House.

Hodgson, G. M. (1993). *Economics and evolution: Bringing life back to economics*. Cambridge: Polity Press.

Hodgson, G. M. (1995). *Economics and biology*. Aldershot: Edward Elgar.

Hodgson, G. M. (1999). *Evolution and institutions: On evolutionary economics and the evolution of economics*. Cheltenham: E. Elgar.

Hodgson, G. M. (2001). Is social evolution Lamarckian or Darwinian? In J. Laurent & J. Nightingale (Eds.), *Darwinian and evolutionary economics* (pp. 87–120). Cheltenham: E. Elgar.

Hodgson, G. M. (2004). *Evolution of institutional economics: Agency, structure, and Darwinism in American institutionalism*. London: Routledge.

Humphreys, P. (2016). *Emergence: A philosophical account*. Oxford: Oxford University Press.

Intermann, K. (2015). Distinguishing between legitimate and illegitimate values in climate modeling. *European Journal for Philosophy of Science, 5*, 217–232.

Knuuttila, T., & Loettgers, A. (2016). Models templates within and between disciplines: From magnets to gases — And socio-economic systems. *European Journal for Philosophy of Science, 6*(3), 377–400.

Mainzer, K. (2007). *Thinking in complexity. The computational dynamics of matter, mind, and mankind* (5th ed.). Berlin: Springer.

Mäki, U. (2009). Economics imperialism: Concept and constraints. *Philosophy of the Social Sciences, 39*(3), 351–380.

Mitchell, S. D. (2009). *Unsimple truth: Science, complexity, and policy*. Chicago: The University of Chicago Press.

Morrison, M. (2015). *Reconstructing reality. Models, mathematics, and simulations*. New York: Oxford University Press.

Nelson, R., & Winter, S. (1982). *An evolutionary theory of economic change*. Cambridge, MA: Harvard University Press.

Niiniluoto, I. (1993). The aim and structure of applied research. *Erkenntnis, 38*, 1–21.

Niiniluoto, I. (1995). Approximation in applied science. *Poznan Studies in the Philosophy of the Sciences and the Humanities, 42*, 127–139.

Niiniluoto, I. (2020). Interdisciplinarity from the perspective of critical scientific realism. In W. J. Gonzalez (Ed.), *New approaches to scientific realism* (pp. 231–250). Boston/Berlin: De Gruyter.

Ortega y Gasset, J. ([1933] 1997). *Meditación de la Técnica*, edited by Jaime de Salas and José María Atencia. Madrid: Santillana.

Pies, I., & Leschke, M. (Eds.). (1998). *Gary Beckers ökonomischer Imperialismus*. Tübingen: Mohr Siebeck.

Radnitzky, G., & Bernholz, P. (Eds.). (1987). *Economic imperialism: The economic method applied outside the field of economics*. New York: Paragon House.

Rescher, N. (1988). *Rationality: A philosophical inquiry into the nature and the rationale of reason*. Oxford: Clarendon Press.

Rescher, N. (1998a). *Predicting the future*. New York: State University of New York Press.

Rescher, N. (1998b). *Complexity: A philosophical overview*. New Brunswick: Transaction Publishers.

Rescher, N. (1999). *The Limits of Science* (rev ed.). Pittsburgh: University of Pittsburgh Press.

Rescher, N. (2019). *Philosophical clarifications: Studies illustrating the methodology of philosophical elucidation*. Cham: Palgrave Macmillan.

Rochefort-Maranda, G. (2016). Simplicity and model selection. *European Journal for Philosophy of Science, 6*(2), 261–279.

Sah, R. K., & Stiglitz, J. E. (1986). The architecture of economic systems: Hierarchies and polyarchies. *American Economic Review, 76*(4), 716–727.

Schenk, K.-E. (2006). Complexity of economic structures and emergent properties. *Journal of Evolutionary Economics, 16*, 231–253.

Sen, A. (1986). Prediction and economic theory. In J. Mason, P. Mathias, & J. H. Westcott (Eds.), *Predictability in science and society* (pp. 3–23). London: The Royal Society and The British Academy.

Simon, H. A. ([1973] 1977). The organization of complex systems. In: H. H. Pattee (Ed.), Hierarchy theory (pp. 3–27). New York: G. Braziller. Reprinted in: H. A. Simon, *Models of discovery* (pp. 245-264). Boston: Reidel.

Simon, H. A. ([1990] (1997). Prediction and prescription in systems modeling. *Operations Research, 38*, 7–14. Reprinted in: H. A. Simon, *Models of bounded rationality. Vol. 3: Empirically grounded economic reason* (pp. 115–128). Cambridge, MA: The MIT Press.

Simon, H. A. (1996). *The sciences of the artificial* (3rd ed.). Cambridge, MA: The MIT Press.

Simon, H. A. (2001). Complex systems: The interplay of organizations and markets in contemporary society. *Computational and Mathematical Organization Theory, 7*(2), 79–85.

Spohn, W. (2011). Normativity is the key to the difference between the human and the natural sciences. In D. Dieks, W. J. Gonzalez, S. Hartmann, T. Uebel, & M. Weber (Eds.), *Explanation, prediction, and confirmation* (pp. 241–251). Dordrecht: Springer.

Stigler, G. J. (1984). Economics: The Imperial science? *Scandinavian Journal of Economics, 86*, 301–313.

Thagard, P. (1992). *Conceptual revolutions*. Princeton: Princeton University Press.

The Royal Swedish Academy of Sciences. (2016). *The Nobel Prize in Physics 2016*, https://www.nobelprize.org/nobel_prizes/physics/laureates/2016/press.html. Accessed on 1.12.2016.

Tiropanis, T., Hall, W., Crowcroft, J., Contractor, N., & Tassiulas, L. (2015). Network science, Web science, and Internet science. *Communications of ACM, 58*(8), 76–82.

United Nations. (2015). *Paris Agreement: Framework Convention on Climate Change*, 30 November 2015 to 11 December 2015, Paris, 12.11.2016, https://unfccc.int/resource/docs/2015/cop21/eng/l09r01.pdf. Accessed on 28.11.2016.

Wagner, A. (2012). The role of randomness in Darwinian evolution. *Philosophy of Science, 79*(1), 95–119.

Woodward, J. (2003). *Making things happen*. Oxford: Oxford University Press.

Worrall, J. (1988). The value of a fixed methodology. *The British Journal for the Philosophy of Science, 39*, 263–275.

Worrall, J. (1989a). Fresnel, Poisson and the white spot: The role of successful predictions in the acceptance of scientific theories. In D. Gooding, T. Pinch, & S. Schaffer (Eds.), *The uses of experiment* (pp. 135–157). Cambridge: Cambridge University Press.

Worrall, J. (1989b). Fix it and be damned: A reply to Laudan. *The British Journal for the Philosophy of Science, 40*, 376–388.

Worrall, J. (1998). Realismo, racionalidad y revoluciones. *Ágora, 17*(2), 7–24.

Worrall, J. (2006). Why randomize? Evidence and ethics in clinical trials. In W. J. Gonzalez & J. Alcolea (Eds.), *Contemporary perspectives in philosophy and methodology of science* (pp. 65–82). A Coruña: Netbiblo.

Chapter 3
Multiscale Modeling: Explanation and Emergence

Robert W. Batterman

Abstract I discuss two extreme views about emergence and explanation in the face of the following facts. The continuum equations of fluid mechanics and material behavior (bending of beams) are remarkably successful. They continue to be used for engineering and building purposes despite the fact that they almost completely ignore lower scale structure and detail. In fact, they treat fluids and steel beams as continuous blobs with no lower scale (atomic/molecular) structure whatsoever. The explanatory question is how such equations can be so successful. Extreme reductionists views expect that a bottom-up explanation from the actual fundamental lower scale structure will answer this explanatory question. Extreme emergentist views deny such reductionist pictures and hold that no bottom-up explanation is possible. The paper argues for a more sophisticated analysis that recognizes the importance of structures at scales intermediate between the atomic and the continuum. It is an analysis that appeals to sophisticated mathematics called "homogenization theory." One can, in fact, provide a nonreductive explanation of the relative autonomy of the continuum equations from the lower scale details.

Keywords Models · Multi-scale · Explanation · Emergence · Reduction · Homogenization · Continuum equations

I would like to thank Julia Bursten for helpful comments. This research was supported by a grant from the John Templeton Foundation.

R. W. Batterman (✉)
Department of Philosophy, University of Pittsburgh, Pittsburgh, PA, USA
e-mail: rbatterm@pitt.edu

3.1 Introduction: Fundamental Theories and Autonomy

In 1963 Clifford Truesdell declared that physicists are at a "disadvantage when facing common experience." (Truesdell 1984, p. 47). The reason for this, as he saw it, is that physicists are too focused on *fundamental theories* of the structure of matter. To a large extent they ignore phenomenological theories (such as continuum mechanics) that describe "less fundamental" or phenomenological behaviors of matter at everyday length and time scales. While 1963 was a long time ago, and we have since seen many debates about what should count as fundamental physics,[1] philosophical theorizing mostly still privileges the reductionist approach that Truesdell scorns. Truesdell continues:

> The training of professional physicists today puts them under heavy disadvantage when it comes to understanding physical phenomena, much as did a training in theology some centuries ago. Ignorance commonly vents itself in expressions of contempt. Thus "physics", by definition, is become exclusively the study of the structure of matter, while anyone who considers physical phenomena on a supermolecular scale is kicked aside as not being a "real" physicist. Often "real" physicists let it be known that all gross phenomena easily could be described and predicted perfectly well by structural theories; that aside from the lack of "fundamental" (i.e. structural) interest in all things concerning ordinary materials such as water, air, and wood, the blocks to a truly "physical" (i.e. structural) treatment are "only mathematical". (Truesdell 1984, p. 47)

The idea that "all gross phenomena easily could be described and predicted perfectly well by structural theories" is one of the core aspects of reductionism. The idea that the only impediment to such reductionist predictions is mathematical serves to highlight the fact that reductionists often dismiss mathematical difficulties as *merely pragmatic* and that such pragmatic problems do not stand in the way of *in principle* derivations from fundamental theory.

In the next section, I discuss some "old" theories that are still around today and that remain remarkably successful. These are continuum theories of matter that are used to describe, predict, and explain the behaviors of materials such as steel beams (how they bend under loads) and fluids (how they flow through pipes). They are quite remarkable in that we still employ them in our engineering projects, yet they make absolutely no reference to the fact that steel and water, for example, are composed of atoms and molecules. An important question of interest is how they can be so useful despite not being "fundamental" or "structural." In Sect. 3.3 I elaborate on this question, framing it as an explanatory request concerning the *autonomy* of continuum scale theories from atomic and molecular details. Some philosophers, notably Jerry Fodor, have held that such autonomy is inexplicable, leading to the view that autonomous theories might reasonably be considered to be emergent. In Sect. 3.4 I argue that the autonomy is genuine but that it can, in fact,

---

[1]Phillip Anderson's famous paper, "More is Different," represents the beginning of a backlash against such reductionist thinking at least by physicists (Anderson 1972).

be explained. However, the explanation is *not* one that would appeal to Truesdell's "real" physicists.

3.2 Old Theories

The nineteenth-century saw the development of a number of physical theories describing the behavior of different forms of matter. As examples we can take fluid mechanics with its fundamental equation known as the *Navier-Stokes* equation, and continuum mechanics including the equation for the behavior of bending rods and beams known as the *Navier-Cauchy* equation. Both of these theories treat matter as a continuum. That is to say, there is no mention of any structure whatsoever at any scale. In effect the equations assume that as one zooms in on a fluid or solid, no matter how much magnification one employs, one will not see any structure. The materials are assumed to be completely homogeneous all the way down to the scale of infinitesimals.

Let us consider the Navier-Cauchy equation for elastic solids:

$$(\lambda + \mu)\nabla(\nabla \cdot \mathbf{u}) + \rho\nabla^2\mathbf{u} + \mathbf{f} = 0.^2 \qquad (3.1)$$

The *material* parameters, λ and μ are the "Lamé" parameters and are related to Young's modulus. They characterize the bendiness of the material. When one specifies the values for these material parameters along with the density ρ, one defines the system to which the equation applies. The values for these parameters are determined by experiments in a laboratory: One puts the material in a vise and measures how much it extends upon stretching and how much it shortens upon being compressed.

The remarkable fact about the continuum equations is that they *are still used today* successfully to construct bridges, buildings, and boats. As noted, they do not make any reference to the fact (as we now know) that materials like steel are *not* homogeneous at all scales. How is this possible? *How can such equations be successful and at the same time exhibit quite remarkable autonomy from lower scale details*?

To my mind, this is the critical question that is often ignored in the debate about reduction vs. emergence. The fact of multiple realizability challenges the reductionist to answer the following question:

- (**MR**) How can systems that are heterogeneous at some (typically) micro-scale exhibit the same pattern of behavior at the macro-scale?

$^2\mathbf{u}$ is the displacement vector, ρ is the material density, and $\mathbf{f}$ are the body forces acting on the material.

In the current context, asking how continuum equations can be used successfully to build bridges that support loads, etc. is equivalent to asking (**MR**). After all, we know that these equations allow us to build bridges out of materials (wood, steel, etc.) with differing microstructures.

Philosophical references to emergence often suggest a kind of *autonomy* had by emergent properties of a whole. In the philosophical literature this notion of autonomy is often expressed in terms of the unpredictability and the inexplicability of the properties of the whole in terms of the properties of the parts. Many times too, there is a claim that the emergent whole exhibits causal behavior that is novel and not displayed by the behaviors of its components. Thus one feature of "standard" philosophical accounts of emergence is the inexplicability of the autonomous aspects of the emergents' behaviors. I think this standard view is mistaken. One can have autonomy—understood primarily as a effective lack of dependence on lower scale details—while at the same time strive for an explanatory account of the origin of that autonomy.

The nineteenth-century theories lack the "structural" or "fundamental" status that Truesdell's "real" physicists demand. Nevertheless, as noted, they possess a kind of autonomy from those fundamental theories that do describe the submolecular/atomic structure that we know constitute the fluids, gases, beams, and rods. Despite this autonomy, these phenomenological equations accurately describe the everyday behavior of the materials. Of course, according to the "real" physicists, this autonomy is only apparent, and the only thing blocking a genuine fundamental understanding of the every day behavior is mathematical complexity. The idea is that *in principle*, if it were not for some mathematical difficulties, the fundamental theories would describe the behavior of everyday materials perfectly well.

However, even if the autonomy is only apparent in this way, it remains a *fact about the world and a fact about our theorizing about the world*, that we successfully use such phenomenological theories for our engineering projects. These facts are something that the fundamental theories should be able to explain, if they are truly fundamental. After all, these facts about the successes of continuum modeling are (meta) patterns of real world behavior. An answer to why continuum equations are successful in so many varied instances is a response, once again, to (**MR**).

3.3 Autonomy

In the contemporary physics literature and in popular science the term "emergence" is often used in reference to "protected states of matter" or "protectorates." For example, Laughlin and Pines (2007) use the term "protectorate" to describe domains of physics (states of matter) that are effectively independent of microdetails of high energy/short distance scales. A (quantum) protectorate, according to Laughlin and Pines is "a stable state of matter whose generic low-energy properties are determined by a higher organizing principle and nothing else." (Lauglin and Pines 2007, p. 261). Laughlin and Pines do not say much about what a "higher organizing principle" actually is, though they do refer to spontaneous symmetry breaking in this context. For instance, they consider the existence of sound in a solid as an emergent phenomenon independent of microscopic details:

> It is rather obvious that one does not need to prove the existence of sound in a solid, for it follows from the existence of elastic moduli at long length scales, which in turn follows from the spontaneous breaking of translational and rotation symmetry characteristic of the crystalline state." (Lauglin and Pines 2007, p. 261)

It is important to note that Laughlin and Pines do refer to features that exist at "long length scales." Unfortunately, both the conception of a "higher organizing principle" and what they mean by "follows from" are completely unanalyzed. Nevertheless, these authors do seem to recognize what I take to be a very important feature of emergent phenomena—namely, that they display a kind of autonomy. Details from higher energy (shorter length) scales appear to be largely irrelevant for the emergent behavior at lower energies (larger lengths).[3]

Some philosophers, as well, have made reference to the autonomy of the so-called "special sciences" from lower-level or lower-scale theories such as physics. Jerry Fodor's famous paper "Special Sciences, or the Disunity of Science as a Working Hypothesis" is a case in point.

> [T]here are special sciences not because of the nature of our epistemic relation to the world, but because of the way the world is put together: not all natural kinds (not all the classes of things and events about which there are important, counterfactual supporting generalizations to make) are, or correspond to, physical natural kinds. ... Why, in short, should not the natural kind predicates of the special sciences cross-classify the physical natural kinds? (Fodor 1974, pp. 113–114)

Fodor despairs at providing an answer to this last question. For him it is rhetorical— a statement to the effect that the special sciences can be and are *completely* autonomous from lower-level physical theory, to the extent that no explanatory dots may be connected between them. This is reinforced in a later paper "Special Sciences: Still Autonomous After All These Years." (Fodor 1997). There he says:

> The very *existence* of the special sciences testifies to reliable macrolevel regularities that are realized by mechanisms whose physical substance is quite typically heterogenous ... Damn near everything we know about the world suggests that the unimaginably complicated to-ings and fro-ings of the bits and pieces at the extreme *micro*level manage somehow to converge on stable *macro*level properties.
> On the other hand, the "somehow" really is entirely mysterious .... (Fodor 1997, p. 160–161)

He continues by saying that he doesn't even "know how to *think about* why" there could be such macro-level regularities that are multiply realized by heterogeneous micro-level properties. He doesn't know why, as he puts it, "there is anything except

---

[3]Note that this characterization of the most important feature of emergent properties make *no* reference to parts or wholes. I think this demonstrates that mereological conceptions of emergence are most often mischaracterizations. However, this is would take us beyond the scope of the current paper.

physics." (Fodor 1997, p. 161). Comments such as these lead some to think that emergent properties, if they exist at all, will be very spooky indeed.

It is clear that Fodor holds the special sciences to be completely autonomous. His claim, that this autonomy is entirely mysterious, is a further indication that he thinks an explanation of the autonomy is not going to be forthcoming. So, on the one hand, we have Truesdell's "real" physicist who tells us that "all gross phenomena easily could be described and predicted perfectly well by structural theories" and that the only impediment to doing so is the difficulty of the mathematics. And, on the other hand, we have Fodor telling us, in effect, that such prediction, description, and (on a DN conception) explanation is impossible. As usual in philosophy, when such stark contrasts arise, it is likely that something is wrong with the way the problem has been formulated. Both positions are extreme and the arguments for both positions are crude.

To begin to see this, let us return to the form of the Navier-Cauchy equation and focus on the material parameters—the Lamé constants. As noted, without providing values for these parameters (without providing material specificity) the equation does not tell us anything about the world—it is empty formalism. And, we have seen that to provide such material specificity we most always must refer to empirically determined data. But, given that we *know* that our steel beam is not homogeneous at all scales, we must assume that the actual structure at smaller scales plays some role in determining the actual values for the Lamé parameters for a given beam. Does this further knowledge about atomic structure vindicate the reductionist? Will it supply the *in principle* argument the "real" physicist believes to be available?

It seems to me that if the answer is "no," then one must conclude that the apparent autonomy of continuum scale modeling (from atomic scale details) is an example of a kind of genuine (or absolute) autonomy. If this is the case, we will have gone some way toward justifying Fodor's view of the relations between theories or models at different scales. But, as we will see, Fodor is not fully vindicated. The degree of autonomy is not absolute. Lower scale features enable us to partially explain the existence of such continuum scale protectorates.

3.4   Scales, Emergence, and Steel

Real steel exhibits a wide range of distinct structures or heterogeneities as one zooms in from the macroscopic to the atomic. See Fig. 3.1. At scales of 1 m to 10^{-3} m say, the steel girder appears to be almost completely homogeneous: Zooming in with a small microscope will reveal nothing that looks much different. In fact, there appears to be a kind of local scale invariance here.[4] So for behaviors

---

[4]"Local" in the sense that the invariance holds for scales of several orders of magnitude but fails to hold if we zoom in even further, using x-ray diffraction techniques, for example.

Fig. 3.1 Microstructures of steel

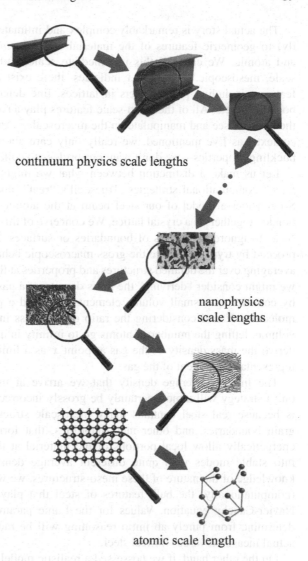

continuum physics scale lengths

nanophysics
scale lengths

atomic scale length

that take place within this range of scales, the steel girder is well-modeled by the Navier-Cauchy equation (3.1).

Now if we jump from this large scale picture to the smallest/atomic scale we see that steel is an alloy that contains both iron and carbon atoms in a highly structured crystalline lattice structure. It appears not at all like the homogeneous beam that exhibits no such structure at all. Somehow between the lowest scale of crystals on a lattice and the scale of millimeters, the low level structure disappears. This at least suggests that the properties of the atomic lattice of steel cannot, by themselves, determine what is responsible for the properties of steel at the macroscale.

The actual story is remarkably complex and intimately involves appeal (primarily) to geometric features of the material at scales *intermediate* between macro and atomic. We can call this intermediate scale, and structural features at that scale,"mesoscopic." As Fig. 3.1 indicates, there exist a host of such mesoscopic features, including point defects in lattices, line defects, slip dislocations, grain boundaries, etc. All of the meso-scale features play a role in the *homogenization* of the steel we see and manipulate on the macroscale.[5] And, naturally, in engineering contexts, as I've mentioned, we really only care about such macro bending and buckling properties—we don't want our bridges to collapse.

Let us make a distinction between what we might call "ab initio" and "*post facto*" computational strategies. Truesdell's "real" physicists endorse the former. So consider a model of our steel beam at the atomic scale where the atoms are bonded together on a crystal lattice. We conceive of this as an infinite perfect lattice so as to ignore the effects of boundaries or surfaces. Ab initio strategies usually proceed by trying to derive the gross macroscopic behavior through some kind of averaging over the detailed structures and properties at the lower scale. For example, we might consider "deriving" the mass density of a gas (as a function of position) by considering a small volume element V around a point x that contains many molecules N and considering the ratio of the mass in the volume element to its volume, letting the number of atoms go to infinity in an appropriate way. We thus derive the mass density of the gas at point x as a limit of a volume average in a representative element of the gas.

The limiting average density that we arrive at using this ab initio (atomic only) strategy will almost certainly be grossly incorrect at continuum scales. This is because real steel contains many mesoscale structures, such as dislocations, grain boundaries, and other meta-stabilities, that form within its bulk and that energetically allow local portions of the material at these higher scales to settle into stable modes with quite different average densities. Until we gain some knowledge of the nature of these meso-structures, we will not be able to determine (computationally) the bulk features of steel that play such a crucial role in the Navier-Cauchy equation. Values for the Lamé parameters that we might try to determine from purely ab initio reasoning will be radically at variance with the actual measured values for real steel.

On the other hand, if we possessed a realistic model of steel at all length scales, then we could conceivably average properly over a representative volume (at a much higher scale than the atomic). But this *post facto* calculation would rely upon data about the system at all scales. Attempts to do this kind of averaging fall under the mathematical theory of *homogenization* which is a kind of multiscale modeling. It is a very different kind of averaging or coarse-graining than that appealed to in the ab initio approach just described. Of course, it needs to be different, since the simple limiting averages, as noted, will almost certainly fail to be empirically adequate.

---

[5] See below for more details about the theory of homogenization.

In broadest outline, the goal of homogenization is to find the

appropriate *homogenized* (or averaged, or macroscopic) governing partial differential equations describing physical processes occurring in heterogeneous materials when the length scale of the heterogeneities tends to zero. In such instances it is desired that the effects of the microstructure reside wholly in the *macroscopic* or *effective* properties via certain weighted averages of the microstructure. (Torquato 2002, p. 305)

Of course the volume averages of different microstructures play some role determining the effective properties of materials. However, it often turns out that the most important contribution to these effective properties is due to geometric or topological features of the heterogeneous microstructures. Here is a simple example that shows how topology can be extremely important. Consider a composite material consisting of *equal* volumes of two materials one of which is a good electrical conductor and one of which is not. A couple of possible configurations are shown in Fig. 3.2.

Let the dark phase be a good conductor and the light phase be a good insulator. Suppose we average over the volumes according to the ab initio strategy discussed above. Then, because the volume fractions of each phase are equal, we would grossly underestimate the bulk conductivity of the material in the left configuration and grossly underestimate its bulk insulating capacity in the right configuration. In this example, the microstructural feature that is most important with respect to the macroscale behavior of the material is *the topological connectedness* of one phase vs. that of the other.

One might object that all one needs to do to save the simple averaging methodology would be to properly weight the contribution of the different phases to the overall average. But this is not something that one can do a priori or through ab initio calculations appealing to details and properties of the individual atoms at the atomic scale.

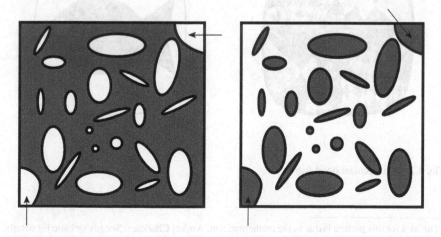

Fig. 3.2 50–50 volume mixture

So in order to *upscale*—in order to determine the macroscopic features of the steel bar (considered now as a composite of various materials with distinct properties)—we need to pay attention to a host of mesoscopic features in addition to the simple volume averages of the different materials making up the composite. These features include "surface areas of interfaces, orientations, sizes, shapes, spatial distributions of the phase domains; connectivity of the phases; etc." (Torquato, 2002, p. 12). In trying to bridge the scales between the atomic domain and that of the macroscale, one needs to connect rapidly varying local functions of the different phases to differential equations characterizing the system at much larger scales. Homogenization theory accomplishes this by taking limits in which the local length (small length scale) of the heterogeneities approaches zero in a way that preserves (and incorporates) the topological and geometric features of the microstructures.

The idea is to try to replace the complicated actual structure of the steel with a simple model that accurately gets the continuum properties right. A nice conception of this is illustrated in Fig. 3.3.[6] The very complicated Escher structure exhibits complexity and heterogeneity at many scales. The goal of homogenization is to eliminate such complexity in a way that the most important features (again, largely geometric) are preserved and become encoded in the values for the material parameters (such as the Lamé parameters).

Most simply, and abstractly, homogenization theory considers systems at two scales: ξ, a macroscopic scale characterizing the system size, and a mesoscopic scale, a, associated with the mesoscale heterogeneities. There may also be applied external fields that operate at yet a third scale Λ. If the mesoscale, a, is comparable with either ξ or Λ, then the modeler is stuck trying to solve equations at that smallest

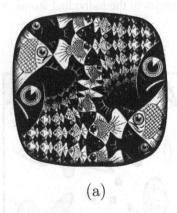
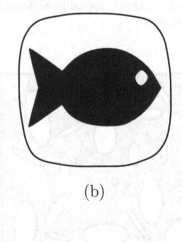

(a)   (b)

Fig. 3.3 Homogenization of fish

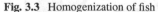

[6]The idea for this picture is due to the mathematician, Andrej Cherkaev. See his website for details. http://www.math.utah.edu/~cherk/.

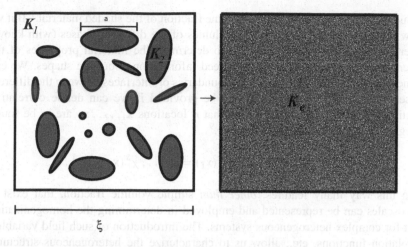

Fig. 3.4 Homogenization limit. (After Torquato 2002, pp. 2, 305–306)

scale. However, as is often the case, if $a \ll \Lambda \ll \xi$, then one can introduce a parameter

$$\epsilon = \frac{a}{\xi}$$

that is associated with the fluctuations at the mesoscale of the heterogeneities—the local properties (Torquato 2002, pp. 305–306). In effect, then one looks at a family of functions u_ϵ and searches for a limit $u = \lim_{\epsilon \to 0} u_\epsilon$ that tells us what the effective properties of the material will be at the macroscale.

Figure 3.4 illustrates this. The left box shows the two scales a and ξ with two phases of the material K_1 and K_2. The homogenization limit enables one to treat the heterogenous system at scale a as a homogenous system at scale ξ with an effective material property represented by K_e. For an elastic solid like the steel beam, K_e would be the effective stiffness tensor and is related experimentally to the Lamé parameters.

An essential element of homogenization or upscaling involves describing the mesoscale structures in a way that connects them to the phenomenological characteristics displayed at the continuum scales. For example, in the left image in Fig. 3.4 we can define indicator or characteristic functions for the different phases as a function of spatial location (Torquato 2002, pp. 24–25). If the shaded phase occupies a region U_s, then an indicator function of that phase is given by

$$\chi^s(\mathbf{x}) = \begin{cases} 1, & \text{if} \quad \mathbf{x} \in U_s \\ 0, & \text{otherwise} \end{cases}$$

Such a function will delimit the volume fraction of the shaded material.[7] But we know that knowledge of the relative volumes of the different phases (with known material parameters) is not sufficient to determine the material properties of the system at continuum scales. We also need information about the shapes. We can define new indicator functions for the boundaries or interfaces between the different phases.[8] Much more information will be provided if we can define correlation functions expressing the probabilities that n locations $x_1, \ldots, x_n$ are to be found in regions occupied by the shaded phase.[9]

$$S_n^s(x_1, \ldots, x_n) = Pr\left\{\chi^s(x_1) = 1, \ldots, \chi^s(x_n = 1)\right\}.$$

In this way many features, *other than* simple volume fraction, that exist at microscales can be represented and employed in determining the homogenization limit for complex heterogeneous systems. The introduction of such field variables, correlation functions, etc., allow us to characterize the heterogeneous structures above the atomic scales. Homogenization theory then can exploit these features and determine (ranges of) values for the phenomenological parameters at the continuum stage.

The main lesson to take from this very brief discussion is that the physics at these mesoscopic scales needs to be invoked if one is to be able to explain macroscale behavior, and if we are to answer (**MR**). Bottom-up (ab initio) modeling of systems that exist across a large range of scales is not sufficient to yield observed properties of those systems at higher scales. Neither is complete top-down modeling. After all, we know that the parameters appearing in continuum models must depend upon details at lower scale levels. The interplay between the two strategies—a kind of mutual adjustment in which lower scale physics informs upper scale models and upper scale physics corrects lower scale models—is complex, fascinating, and unavoidable.

3.5 Conclusion

We have seen two extreme views about emergence and explanation. On the one hand, Truesdell's "real" physicists insist that, but for some mathematical unpleasantness, one can *in principle*, ab initio derive all "gross phenomena" or the continuum behaviors of materials from fundamental atomic/structural theories. On the other hand, Fodor insists that the special sciences are autonomous from

---

[7]And, clearly, in a two phase system such as that in the figure, the complement of this function will delimit the volume fraction of the other phase.

[8]These will be generalized distribution functions.

[9]See Torquato (2002) for a detailed development of this approach.

lower level theories/details and that *no* bottom-up explanation can be offered from fundamental/structural theories for the very existence of that autonomy.

Both sorts of views, however, must confront the same facts. And, these facts include the remarkable successes and stability of continuum scale modeling. These models are successful despite the fact that they pay virtually no attention to lowest scale structures and details. If the "real" physicists are right, then they should be able to explain this success in the bottom-up fashion they insist is always possible. But we have seen, in Sect. 3.4 that such explanations are doomed to fail. Nevertheless, one can explain the success and autonomy of the continuum modeling. The explanation, provided by the theory of homogenization (and related techniques such as the renormalization group) is neither purely bottom-up nor purely top-down.

The truth about emergence (or at least about one very real and important aspect of emergence—autonomy) is much more involved and complex than is usually realized. Philosophers and physicists often take all or nothing, extreme, attitudes toward the problem. This paper argues that a more subtle, careful, and sophisticated analysis will enable philosophers to establish and explain the existence of the relevant kind of autonomy displayed by models at higher scales from those at lower scales. In essence, we can provide a non-reductive explanation of this autonomy. This preserves a genuine feature of what many have thought emergence to involve, while at the same time removing much of the "spookiness" often attributed to emergentist thinking.

References

Anderson, P. W. (1972). More is different. *Science, 177*(4047), 393–396.

Fodor, J. (1974). Special sciences, or the disunity of sciences as a working hypothesis. *Synthese, 28*, 97–115.

Fodor, J. (1997). Special sciences: Still autonomous after all these years. *Philosophical Perspectives, 11*, 149–163.

Laughlin, R. B., & Pines, D. (2007). The theory of everything. In M. Bedau & P. Humphreys (Eds.), *Emergence: Contemporary readings in philosophy and science* (pp. 259–268). Boston: The MIT Press.

Torquato, S. (2002). *Random heterogeneous materials: Microstructure and macroscopic properties*. New York: Springer.

Truesdell, C. (1984). *An idiot's fugitive essays on science: Methods, criticism, training, circumstances*. New York: Springer.

Part II
Pragmatist Approaches to Methodology of Science

Part II

Chapter 4
Methodological Pragmatism

Nicholas Rescher

Abstract Pragmatism adopts the procedure of assessing process by product, of evaluating ways of doing things on the basis of their functional efficacy in realizing the collective objectives. This can be done either *directly* by assessing the quality of the product, or *obliquely* by assessing the efficacy of procedures and methods in engendering high quality products. Methodological (as contrasted with Productive) Pragmatism adopts the latter approach. Its prime advantage is its greater reliability in providing for quality control thanks to the superior realism of the evaluations it underwrites.

Keywords Methodological · Pragmatism · Procedures · Methods · Efficacy

4.1 The Basic Idea

Philosophers do not have the term "pragmatism" to themselves. It has a characteristic and long-established usage in relation to political matters. Someone is "pragmatic" in political affairs if they are willing to forego what they might ideally want for the sake of achieving something superior to what they have. In being politically pragmatic one is prepared "to settle for half a loaf" by accepting an achievable improvement rather than insisting upon a maximally desirable outcome. Realizing that the best can become the enemy of the good the pragmatic politician accepts a realizable improvement at the piece of at the cost of forsaking a potentially unattainable ideality.

Where Immanuel Kant sought to explain issues on the basis of general principles regarding the "conditions under which alone" a certain type of cognitive task can *possibly* be accomplished, the pragmatist wants to determine—less by abstract

N. Rescher (✉)
Center for Philosophy of Science, University of Pittsburgh, Pittsburgh, PA, USA
e-mail: rescher@pitt.edu

© Springer Nature Switzerland AG 2020
W. J. Gonzalez (ed.), *Methodological Prospects for Scientific Research*, Synthese Library 430, https://doi.org/10.1007/978-3-030-52500-2_4

analysis than by experiential trial and error—the considerations with which this task can be *efficiently and effectively* accomplished.

Like empiricism or idealism or other philosophical isms, pragmatism encompasses a considerable variety of rather different doctrinal positions. But overall they share certain characteristic conceptions or perspectives, primarily the following:

- that considerations of abstract general principles do not of themselves suffice for a satisfactory resolution of "the big questions" of philosophy and philosophy of science above all.
- that actual praxis—trial and error in the rough and tumble arena of actual experience—affords a salient standard for assessing the adequacy of theoretical doctrines.

The idea is that theory exists for the sake of practice and that working out in matters of application and implementation is a key article of adequacy where mattes of theorizing are at issue.

4.2 Pragmatic Realism

Pragmatism is sometimes said to reject philosophical realism, but one is "realistic" in abandoning idealized and impetus possibilities in favor of beneficial practicalities. He advances by practicable realizable steps rather than risking the mishap of ideologically inspired leaps. He looks to what can be made to work out for the better, rather than to what would ideally be for the best, but is almost certainly unachievable. Philosophical pragmatism is something remotely related but decidedly different. To begin with its orientation is primarily toward matters of belief and conduct rather than political action.

And this perspective has a special bearing on the problem of truth. For as pragmatists see it we have no direct and epistemologically unmediated way of getting at the truth of things without the detour of what we think to be so. They have it that the instruction: "Please tell me what is true directly and immediately without any reference to what you think there is good reason to consider as such" is beyond prospect of implementation. As William James picturesquely put it, "the trail of the human serpent" runs across all of these philosophical matters—truth, justice, beauty, and the rest.

And so, philosophical pragmatism has one salient feature in common with the political sense of the term, namely a disinclination to rely on plausible-seeming general principles and abstract idealizations but instead it looks to "what works" and to prioritize successful application in some section of practice. For like the political pragmatist, the philosophical pragmatism is prepared to forsake general principles, doctrinal ideologies and theoretical idealizations and use instead as this guide the arbitrariment of experience. Its focal concern is for outcomes, for how things evaluate in practice.

The pragmatist does not propose to proceed on the basis of idiosyncratically appealing general principles. Especially in matters of cognition and communication he has confidence in the established practice of community. Seeing that in these matters we here arrived at what we are by luck and error in school of better experience and, as rational beings, have inclined toward the quasi-Darwin's process of rational evaluation by preferring the fittest practices to remain. There is a strong pro-perspective in favor of the established ways. (Only in matters of reformist practitioners have the main philosophical pragmatists evolved toward a different and less conscientious view. Doctrinal consistency has been one of those theoretical general principles toward which pragmatists have always displayed always sceptical distrust.)

Display 1	
Key perspectives of pragmatism.	
DE-EMPHASIZE	EMPHASIZE
theory	experience and practice
a priori insights (eternal verities)	experiential lessons
general principles	rules of practice
subjective centralities	general consensus
intuition	trial and error
ideology	common sense

Like the utilitarian, the pragmatist is interested in results. But then the utilitarian focusses on the promotion of happiness, the pragmatist looks more broadly to functional efficacy at large. But just like the utilitarian cousin for "the greatest happiness *of the greatest number*," the pragmatist takes a line that is publically and communally oriented. Unlike the Aristotelian concern for personal well-being (*endaimonia*), the pragmatist looks for confirmation to the community at large. (The exception which "provides the rule" here was William James who in his concern for the psychology of the individual—especially in relation to matters of faith and belief—gave pragmatism a theological and personalists turn that horrified C. S. Peirce.)

As Display 1 serves to indicate, philosophical pragmatism has two doctrinal sides: one negative and one positive.

On the negative side, pragmatism rejects the idea of validating beliefs (and especially philosophical beliefs) on the basis of general principles. Kantian a priorism and the resource of transcendental arguments to conditions under which alone certain objectives can be achieved are anathema to the pragmatist. He will have no truck with prospective reasoning from self-evolved fundamentals to substantial conclusions.

On the positive side endorses the recourse to trial and error and the arbitrament of experience. He is prepared to endorse beliefs which prove successful in the course of their application and implementation. Working out successfully in actual employment is the crux of the merit. Here the validation proceeds retrospectively

through what emerges unscathed from the fiery furnace of trial and error. In matters
of procedural validation, pragmatists are willing to "wait and see." They reject the
idea of legitimation on the basis of intuitively accessible general principles. The
pragmatism's motto is then "the proof of the pudding lies in the eating." With regard
to the ancient controversy in Greek medicine between the dogmatic theorist and
the sceptical empiricists, philosophical pragmatism stands squarely within the latter
camp. It proposes to let experience be the teacher and commits to principle of "wait
and see." Dogmatic prejudgments are anathema to the pragmatism.

4.3   Methodological Pragmatism

The methodological pragmatism at issue here differs from other versions of the
doctrine in three principal respects:

1. *Criteriology*. Practical efficacy by way of successful implementation is not to be
   taken as part of the meaning of truth—let alone of its definition—but is rather
   a feature of verification and criteriology of test rather than definition. This shift
   is made to avert the objection that *defining* truth in terms of utility deflects the
   concept away from anything like its traditional conception.
2. *Methodology*. Pragmatic considerations are now to be brought to bear statistically
   in general methods for claim-validation rather than directly and immediately
   upon claims themselves. After all, the natural test of anything instrumental—
   methods, procedures, tools—is that of efficacy with respect to the task at issue,
   and one of the key aims of factual knowledge is its role in the effective gridwork
   of action. This shift from claims to methods of claim verification is made to avert
   the objection that endorsement of false or problematic claims can often provide
   for successful operation.
3. *Impersonality* and *Objectivism*. It is a central idea of methodological pragmatism
   that the success at issue in the acceptance (implementation) of a belief is a
   matter not of its contribution to the psychic well-being of its expert but rather
   of the extent to which the use of this belief in the guidance of action is efficient
   and effective overall—for anyone, anywhere, and on any occasion. It is not a
   matter of the utility the believer's success through *holding* that belief but of the
   impersonal utility of that claim in matters of application and implementation
   at large. This shift is protected by the many objections against James' version
   of pragmatism on grounds that its enmeshment in the subjectivism of personal
   psychology. Accordingly, the objectivistic methodological pragmatism I have
   expounded differs in various crucial and decisive respects of the subjectivistic
   thesis pragmatism espoused by William James. It substantially returns the subject
   to the condition adopted by pragmatism's founder, C. S. Peirce, for whom
   pragmatic considerations spoke on behalf of methods of truth-determination—
   in particular the scientific method—viewing particular claims and contentions as
   acceptable only insofar as pragmatically validated methods spoke on their behalf.

Traditional pragmatism saw the connection between the truth of a contention and the utility of accepting it as a basis for belief and action were established directly and immediately: if it underwrites successful action it merits acceptance as true. Methodological pragmatism takes an entirely different line. Its procedure is to determine the acceptability of beliefs by means of the methods employed for their verification, and to apply the standard of utility-of-results only at the synoptic and statistical level of these methods. It sees the *appropriate* method as the one which does best at determining beliefs when application and implementation are best on the whole, and then takes these methods as our (best available) means for truth estimation. In consequence the link between applicative efficacy and truth is no longer seen as direct, but is methodologically mediated.[1]

4.4 Pragmatic Appropriateness and Cognition

The core and crux of pragmatic validation lies in its taking a functionalistic perspective. Its validating *modus operandi* proceeds with reference to the aims and ends of whatever happens to be the enterprise at issue. The aim of the enterprise of inquiry is to get answers to our questions. And not just answers but answers that can warrantedly be seen as being appropriate through success in matters of explanation and application. And so on pragmatic grounds the rational thing to do in matters of inquiry is to adopt that policy which is encapsulated in the idea that answers to a question for which one need/want an answer—for which the available evidence speaks most strongly is to be accepted until such time as something better comes along.

In line with this perspective, a realistic pragmatism insists upon pressing the question: "If *A* were indeed the correct answer to a question *Q* of ours, what sort of evidence could we possibly obtain for this?" And when we actually obtain such evidence—or at least as much of it as we can reasonably be expected to achieve—then pragmatism enjoins us to see this as sufficient. ("Be prepared to regard the best that can be done as good enough" is one of pragmatism's fundamental axioms.) If it looks like a duck, waddles like a duck, quacks like a duck, and so on, then, so pragmatism insists, we are perfectly entitled to stake the claim that it is a duck—at any rate until such time as clear indications to the contrary come to light. Once the question "Well what more could you reasonably ask for?" meets with no more than hesitant mumbling, then sensible pragmatists say: "Feel free to go ahead and make the claim." While the available information is all too incomplete and imperfect (as fallibilism cogently maintains), nevertheless, in matters of inquiry (of seeking for answers to our questions) we can never do better than to accept that answer for which the available evidence speaks most strongly—or at least to do so until such time as something better comes along.

---

[1] For further detail regarding this position see the author's *Methodological Pragmatism* (1977).

It is not that truth *means* warranted assertability, or that warranted assertability *guarantees* truth. What is the case, rather, is that evidence here means "evidence *for truth*" and (methodologically) warranted assertability means "warrantedly assertable *as true*." After all, estimation here is a matter of truth-estimation, and where the conditions for rational estimation are satisfied we are—ipso facto— entitled to let that estimates stand surrogate to the truth. And in these contexts there is no point in asking for the impossible. The very idea that the best we can do is not good enough for all relevantly reasonable purposes is—so pragmatism and common sense alike insist—simply absurd, a thing of unreasonable hyperbole. Whatever theoretical gap there may be between warrant and truth is something which the very nature of concepts like "evidence" and "rational warrant" and "estimation" authorizes us in crossing.

And so at this point we have in hand the means for resolving the question of the connection between thought and reality that is at issue with "the truth." The mediating linkage is supplied by heeding the *modus operandi* of inquiry. For cognition is a matter of *truth estimation*, and a properly effected estimate is, by its nature as such, something that is entitled to serve, at least for the time being and until further notice, as a rationally authorized surrogate for whatever it is that it is an estimate of.

Consider the following dialogic exchange:

Q: Why should we adopt the policy at issue?
A: Because it is the best one can do in the circumstances.
Q: But why shall I regard the best I can do as good enough?
A: Well, it certainly is not necessarily correct. But the fact remains that it is the best one can do, and that is all that you can (rationally) call for.
Q: But is this line of reasoning not circular? Are you not in effect insisting for its validation that very policy whose validation is in question?
A: That's true enough. But that's exactly how matters should be.
Q: How can you claim this? Is the argumentation not improper on grounds of self-innovation and self-reliance—that is, on grounds of vicious circularity?
A: No. The circularity is there alright. But there is nothing vicious about it. IT is self-supportive and thus is exactly what a thoroughly rational mode of validation should be. For where rationality is involved, self-supportingness is a good thing and circularity is not only unproblematic but desirable. Who would want a defense of reason that is not itself reasonable? Reason and rationality not only can but must be called upon to speak upon their own behalf.

Thus insofar as inquiry into the nature of the real is a matter of truth estimation, the process at issue is and must be one that enjoys reason's seal of approval. For of course rational acceptance cannot be random, fortuitous, haphazard; it must be done in line with rules and regulations, with programs and policies attuned to the prospects of realizing the objectives inherent in the situation at hand.

4.5   The Primacy of Method

Deflationary epistemologists are fearful that if we take a hard objectivistic line on the meaning of truth, then truth becomes transcendentally inaccessible and scepticism looms. And they accordingly insist that we soften up our understanding of the nature of truth. But another option is perfectly open, namely to retain the classical (hard) construction of the meaning of truth as actual facticity ("correspondence to fact") and to soften matters up on the epistemological/ontological side by adopting a "realistic" view of what is criteriologically required for staking rationally appropriate truth claims.

Pragmatists accordingly have the option of approaching "the truth" with a view to the methodology of evidence—of criteriology rather than definitional revisionism. The sort of truth pragmatism that moves in this (surely sensible) direction is one that does not use pragmatic considerations to validate claims and theses directly, but rather uses inquiry methods (claim-validating processes) for this purpose, while validating these practices themselves not in terms of the truth of the products (a clearly circular procedure) but in terms of the capacity of their products to provide the materials for successful prediction and effective applicative control. Accordingly, the most promising position here is—as I see it—a methodological pragmatism rather than a thesis pragmatism. That is, it is a position that assesses thesis assertability in terms of the methodological processes of substantiation and their assesses method appropriateness in terms of the practical and applicative utility—systematically considered—of the thesis from which the methods vouch. Such an approach calls for a prime emphasis on the methodology of truth-estimation, bringing into the forefront the processes of evidentiation and substantiation by which we in practice go about determining what to accept as truth.

We live is a complex, varied, and changeable world. And as we go about pursuing our needs and wants we cannot expect its free and easy cooperation, and certainty cannot expect systematic success when using inappropriate and ineffective methods in the pursuit of our cognitive and practical purposes. Granted, here and there a generally unsuitable modus operandi may by luck and chance yield a descript result but this must be accounted a haphazard fluke. On the whole—at the level of a statistical systematicity—the capacity of a method to deliver desired results must be accounted as the surest and most telling sign that we can have. And this general fact obtains in particular with respect to our cognitive methods and procedures.

Cognition, after all, is a kind of tracking game designed to bring our cognitive models of existence into accord with "the real thing." It is a game against nature (or perhaps more properly *with*) it, and while here—much as at chess or cards— following generally inappropriate rules may occasionally yield a good result. This is something that is just not going to happen at the level of statistical regulating. And with cognition just as with games systemic success in securing results that "work out to satisfaction" is the surest available sign of methodological adequacy.

Still, just how reliable are the truth-estimates that we can manage to get on such a basis? This, clearly, is not the place to write a manual on the epistemology of truth estimation. But three telegraphically brief observations should suffice for present purposes.

1. Our confidence in the acceptability of a truth estimate varies immensely with its precision. We might be tempted to squabble about the claim that yonder person is 3.735 meters tall. But the truth of the thesis that his height is between 1 meter and 4 meters is beyond (reasonable) question.
2. This trade-off between precision and tenability means that our comparatively imprecise claims about every-day life matters are less science than its presence and highly general claims of natural science. The truth of the claim in science at the theoretical future is not as such as is the truth of claims like "The population of New York exceeds six million."
3. With those complex issues at the theoretical frontier of science we are well advised to speak not of unqualified truth as such, but rather of our "best-estimates" of the truth as we are able to realize them with the investigative resources at our disposal. (The common sense realism of everyday-life matters is thus on securer ground than a scientific realism which claims that the objects of scientific inquiry exist in just exactly the descriptive manner in which present-day science conceives of them.)

And so it is on this pragmatic basis that a methodological approach is able to provide a plausible rationale for our confidence on the determinacy of the scientific method as we in practice have it.

The most promising approach to the problem of truth-claim validation would accordingly be to focus on the epistemology of truth estimation and to leave the matter of its definition alone, allowing this to be addressed via the classic conception of truth as *adaequatio ad rem*, as correspondence with (mind-independent) reality. After all, no useful purpose is ever achieved by attributions of "absolute (or "ultimate") truth" or "absolute (or "ultimate") reality in matters of concrete detail." Where plain "truth" and "reality" will not serve, nothing will. Truth can accordingly be left to enjoy the "transcendental" construction that is has always enjoyed. To be sure, the matter of its accessibility is something else again. But this something can be resolved through epistemic deliberations, via the idea of truth-estimation pretty much as standardly conducted.

4.6 Methodological Pragmatism and Levels of Reality

The fact that methodological pragmatism assesses the merit of procedures for truth estimation in terms of the applicative utility of their products has significant theoretical implications. For it is clear that such a proceeding will have to vary with the nature of the objects at issue and to reflect the particular range of phenomena being investigated.

Reality is so complex and many-faceted that we are bound to treat it in different frameworks of consideration that address its very different aspects of its nature. Much will depend on whether the domain at issue is physics, or chemistry, or biology, or economics, or social affairs. The extent to which descriptive precision, phenomenal predictability, and explanatory generality can be realized will differ from field to field depending on the ontology and phenomenology of the domain. In respect to the key issues of pragmatic efficacy as a standard of procedural adequacy there will this have to be a suitable coordination between the levels of reality under consideration.

Thus the statistical stabilities of quantum phenomena exhibit vastly greater stability than those of macro-economics. In most regards the natural world exhibits greater stability than the social, let alone cognitive. The inorganic chemistry of a century ago can still do us better service than its macro-economies.

Accordingly, the sort and extent of the applicative efficacy at issue and the standards and criteria by which they can reasonably be appraised will have to be topic-relative and accordingly differ with the nature of the objects being investigated. In asking we can look for prediction but not control, in economic arrangements we can look for control but not necessarily prediction (owing to the prominence of unforeseeable consequences). As Aristotle wisely observed long ago, "It is the mark of an educated mind to seek in each realm only as much precision as the nature of the subject permits" (349 BC [1946], I, 3 [1094b 24–25]).

In this context the factor of constructive complexity as mirrored in the course of cosmological development plays a crucial role. In the first nanosecond of the universe after the big bang the material for solid state physics had not yet developed; in the first year of cosmic history, there was no material for biology; in the first millennia no need for sociology. In sum, the subject matter indicated for various sciences evolved over time, on the order.

- quantum theory
- solid state physics
- inorganic chemistry
- organic chemistry
- biology
- sociology
- economics

From the angle of methodology each of these domains of phenomenal actuality sets the standard of procedural efficacy at a different and characteristic level of demandingness and precision.

And so in matters, say, of the physics of the unobservable small and that of the mechanics of solid state macro-objects there will have to be difference in the levels of detail that can reasonably be improved upon judgement of methodological efficacy.

4.7 The Impact of Complexity

In matters of methodology, our procedures must always be attuned to the details of the objectives at issue. Throughout the progress of science, technology, and human artifice generally, complexity is self-potentiating because it engenders complications on the side of substantive issues that can only be addressed adequately through further complication on the methodological side of process and procedure. The increase in technical sophistication confronts us with a dynamic feedback interaction between problems and solutions that ultimately transforms each successive solution into a generator of new problems. And these feedback effects operate in such a way that to all intents and purposes the growth rate of the problem domain continually outpaces that of our capacity to produce solutions. Both the problems and the solutions grow more complex in the wake of technological progress, but the crux of the matter lies in the comparatively greater pace of the increase in problem complexity.

However, there yet remains the crucial question of comparative pace. With technological progress, which grows the faster, the manifold of problems to be resolved or the reach and power of our instrumentalities of problem resolution? Now here it might seem that complex technology gives the advantage to problem resolution. After all, do not the cognitive resources that computers provide offset the problems raised by increasing of complexity? Alas, not really.

First of all, it has to be recognized that computers help principally with information *processing* and do not equally address the problems information *acquisition*. And in the course of technological progress these become even more extensive and even more significant. Here the classic dictum holds good: as far as the efficacy of computational information manipulation is concerned, garbage in, garbage out. Moreover, the fact remains that computers do just exactly what they are programmed to do. The level of complexity management they are able to achieve is determined through—and thus limited by—the levels of ingenuity and conceptual adequacy of their programming. No central bank places unalloyed confidence in its economic models. And there is also the problem of unforeseen and unforeseeable interactions within the interact fabric of the operating processes. These "bugs" can result in malfunctions in computer operation even as they can produce accidents in other sorts of systems. And the more elaborate and complex our programs get—particularly in areas where novelty and innovation are the order of the day—the larger the prospect and chances for such mishaps.[2] In every area of applied theorizing, maiden voyages are notoriously fertile in bringing unanticipated difficulties to light.

Granted, computer automated problem solving is one of the wonders of the age. Computers fly planes, land rocket modules on the moon, win chess competitions, develop mathematical proofs. All the same, we have to come face to face here with what might be called a *Hydra effect* after the mythological monster who managed

---

[2]The dramatic failure of the U. S. Department of Internal Revenue's effort to computerize its operations is one particularly vivid example of this.

to grow several heads to take the place for each one that was cut off. The fact is that there is a feedback symbiosis between problems and solutions which operates in such a way that the growth of the former systematically outpaces that of the latter. Accordingly, those sophisticated information and control technologies not so much resolve problems of complexity as enlarge this domain by engendering complexity problems of their own. Despite the enormous advantages that they furnish to intellectual efforts at complexity management, computers nevertheless do not and cannot eliminate but only displace and magnify the difficulties that we encounter throughout this sphere.

The long and short of it is that complexity management via computers will not remove the obstacles to managerial effectiveness exactly because complexity raises problems faster than it provides means for their solution. For technological progress engenders what might be characterized as the "rolling snowball effect" because complexity breeds more complexity through engendering problem-situations from which only additional technical capacity can manage to extract us. Our technological capacity in this regard must grow by exponential leaps and bounds, passing through successively ever more demanding stages. Eventually, no doubt, there will be an end of the line here, for it seems to be a law of nature that all exponential growth must ultimately come to an end. And so in the final reckoning our ability to manage ever greater complexity will eventually become saturated.[3] It is a key teaching of realistic pragmatism that even the best of methods have their limitations.

4.8 Conclusion

So how can one have rational confidence that a methodological approach actually gets at the real truth of things—how can we tell that our truth-estimates are actually good estimates. Here the pragmatically appropriate response will have to run somewhat as follows: "Because those method-provided claims are provided by procedures which yield results that work. They emerge from the use of inquiry methods whose products can be implemented successfully in practice—with success monitored in the usual way of effective application and prediction." For in the complex world that we inhabit we cannot expect systemically successful guidance from faulty cognitive processes. The more varied and complex the range of phenomena being addressed successfully the greater will be our confidence in the adequacy of the methods and consequently in the reliability of their deliverances. And on this basis methodological mediation becomes the gateway to realism, with the acceptability of factual claims vouched for by the efficacy of their methodological procedures.[4]

---

[3]On these issues see the author's *Complexity* (1998).

[4]I am grateful to Professor Wenceslao J. Gonzalez for constructive suggestions in developing this discussion.

References

Aristotle (349 BC [1946]). *Nichomachean ethics*. In: W. D. Ross (Trans. & Ed.), New York: Oxford
    University Press.
Rescher, N. (1977). *Methodological pragmatism*. Oxford: Blackwell.
Rescher, N. (1998). *Complexity*. New Brunswick/London: Transaction Publishers.

Chapter 5
Methodological Incidence of the Realms of Reality: Prediction and Complexity

Amanda Guillan

Abstract The different varieties of methodological universalism have important limitations. These limitations are especially clear when this problem is seen from the ontology of science as the support for both epistemology and methodology of science. Thus, in this paper, the analysis is focused on the problems posed by the methodological universalism to scientific prediction, where the perspective of complexity has a key role. To do this, two main steps are followed. First, the limits to methodological universalism are seen in accordance with the realms of reality. This involves paying attention to the specific features of the natural, social, and artificial phenomena. Second, the research is developed from the perspective of complexity (structural and dynamic), and the methodological relevance of the modes of complexity (above all, epistemological and ontological) and the perspective of historicity are analyzed.

Keywords Methodological universalism · Scientific prediction · Complexity · Historicity

5.1 Introduction

When the problem of scientific prediction is addressed from a methodological perspective, there are, in principle, three levels of analysis (Gonzalez 2012a, 158): (i) the attention to science in general, where the empirical sciences are considered; (ii) the study focused on a group of sciences (natural sciences, social sciences or

This paper was initially prepared for the *Programa FPU* of the Spanish Government. The new version of the paper is related to the research project FFI2016-79728-P supported by the Spanish Ministry of Economics, Industry and Competitiveness (AEI).

A. Guillan (✉)
Center for Resesarch in Philosophy of Science and Technology, Faculty of Humanities and Information Science, University of A Coruña, Campus of Ferrol, Ferrol, Spain

© Springer Nature Switzerland AG 2020
W. J. Gonzalez (ed.), *Methodological Prospects for Scientific Research*, Synthese Library 430, https://doi.org/10.1007/978-3-030-52500-2_5

81

the sciences of the artificial); and (iii) the analysis of a concrete science (such as biology, pharmacology or economics). Proposals of methodological universalism can appear within each one of these levels. This involves conceiving methods which have general validity (or even a method that encompasses a scientific level). Usually, there are approaches that assume that there are some *dominant processes* to develop scientific research and that those dominant processes are valid for one of the aforementioned levels: science, a group of sciences or a concrete science.

Nevertheless, the different varieties of methodological universalism have—in my judgment—important limitations that are especially clear when this problem is seen with regard to the ontological support of the methodology of science.[1] Thus, in this paper, the analysis is focused on the problems posed by methodological universalism to scientific prediction, where the perspective of complexity has a key role.[2] To do this, two main steps are followed. First, the limits to methodological universalism are seen in accordance with the realms of reality. This involves paying attention to the specific features of the natural, social, and artificial phenomena. Second, the research is developed from the perspective of complexity (structural and dynamic), and the methodological relevance of the modes of complexity (above all, epistemological and ontological) is analyzed.

It is possible to think of a third step: the level of reality that is studied, that can be micro, meso or macro. Usually, predictions are in one of these levels of reality, but without emphasizing, in principle (except for cases such as microeconomics and macroeconomics), that prediction is in one level or another (micro, meso or macro). This can become very relevant, but it is not usual to stress the scale of the reality in the dominant methodological approaches, except for certain disciplines and specific theories within them. However, it is an issue that can become really relevant when scientific predictions and prescriptions are analyzed.[3]

5.2 Limits to Methodological Universalism According to the Realms of Reality

When the limits to methodological universalism are considered, as well as their incidence on scientific prediction, a starting point can be the differentiation between natural reality, social realm, and the artificial field. In this regard, Nicholas Rescher has developed an approach where ontology is the support for epistemology and methodology of science, since reality influences both the achievable knowledge and the scientific processes (see Rescher 1998a, 2006, 2010). This is clear in the case of

---

[1]On the different varieties of methodological universalism, see Gonzalez (2012a, 155–175; especially, 155–162).

[2]This chapter draws on and further develops some points originally presented in Guillan (2017, ch. 7, 219–250).

[3]See, in this regard, Gonzalez (2015).

scientific prediction. Thus, the *type of predictions* that can be achieved (regarding their reliability, accuracy, precision, etc.) and the characteristics of the process—in order to make predictions, either within a domain or about a concrete phenomenon (trend extrapolation, analogies, predictive models, etc.)—are issues that, to a large extent, depend on the *type of reality* that we want to predict (see Guillan 2017, ch. 6, pp. 185–215).

5.2.1 Scientific Prediction and Natural Phenomena

Regarding the varieties of methodological universalism, three levels of methodological analysis should be considered (Gonzalez 2012a, 158). In effect, there are proposals of methodological universalism with regard to (a) science in general; (b) a group of sciences; and (c) specific sciences. All these proposals have to deal with the reality of the phenomena, whose variety and variation involve limits to methodological universalism. This feature involves research into the specific characteristics of the different types of reality (natural, social or artificial), which have epistemological and methodological repercussions on science, in general, and on scientific prediction, in particular.

Nicholas Rescher has developed a thoughtful proposal regarding prediction (Rescher 1978, 1998a, 1999a, b, and 2009). His approach is mainly focused on natural sciences. He takes into account that there are important differences between the phenomena studied by the different natural sciences and also within a specific natural science, since natural phenomena are not homogeneous. These ontological differences affect both the epistemological and methodological levels, and they have clear repercussions on scientific prediction. The characteristics of the methods that scientists use to predict depend, to a large extent, on the type of reality that they want to predict (for example, if it is a volatile or a regular phenomenon) and this, in turn, affects the kind of predictive knowledge that is achieved (i.e, its reliability) [Rescher 1998a, 134–135].

In Rescher's proposal, the possibility of obtaining accurate and reliable predictions is linked with the availability of laws, to the extent that he considers that the laws of nature are the most important kinds of patterns that we have in science (Rescher 1998a, 106). In order to predict, we need to find the patterns that phenomena have shown in the past, and laws of nature can provide us those regularities. Thus, the ontological level is fundamental, because the characteristics and behaviour of phenomena (and the possibility that they can be expressed in the form of scientific laws) say us if prediction is possible, to what extent it is possible, and its degree of accuracy and precision.

From this perspective, it is clear that the characteristics of natural phenomena affect the kind of processes to be used, the expected results of the predictive activity, and the type of predictions that we can finally achieve (for example, predictions in biomedicine are usually based on probabilities, so they have less accuracy and precision than in other domains such as astronomy, where stable patterns are available). This also means that prediction has ontological limits. Among those limits, Rescher highlights anarchy, volatility, chance, chaos, arbitrary choice, and creativity (1998a, pp. 133–156). They can affect prediction in various ways: (a) there can be unpredictability (i.e., the complete impossibility of predicting), mainly when phenomena are anarchic; (b) they can involve non predictability (that is, a current impossibility of making predictions), related to instability of phenomena; and (c) in case that they do not prevent prediction, they can still affect the accuracy and precision of the prediction achieved (for example, in some weather forecasts).[4]

Thus, although some ontological obstacles such as chance or chaos do not prevent necessarily prediction, they affect both the type of method to be used and the type of prediction that can be achieved, and, consequently, the kind of knowledge that prediction provides. In accordance, there are a variety of predictive procedures and methods that Rescher classifies in two main types (1998a, pp. 86–88): (a) estimative or judgmental procedures (those developed on the basis of the personal estimations of the experts), and (b) formal or discursive methods (such as trend extrapolation, the use of analogies, the inference from laws, and the predictive models).[5]

Rescher shows a preference for the formal or discursive methods and, among them, for those processes that, in principle, are the most reliable (inference from laws and predictive models). The variety of predictive procedures and methods points out the difficulties of conceiving methods with general validity in the realm of the natural sciences. These difficulties are even greater when the proposals of methodological universalism are not with regard to a group of sciences (such as natural sciences), but with regard to science in general. Rescher's account of prediction (1998a, p. 202) points out that there are limits with special incidence in the realm of the social sciences (such as arbitrary choice or creativity).

For him, there are more methodological difficulties to prediction in social sciences than in the natural sciences, in accordance with the kind of objects and problems addressed by social research.[6] In effect, the use of estimative or judgmental procedures is more frequent in the social realm, where changes are more variable and discontinuous, than in the natural sciences. Changes in natural sciences; especially, in physics, are usually evolutive changes, and, to the extent that evolution involves some kind of continuity, this makes those phenomena still predictable (Rescher 1998a, p. 148).

---

[4]On the distinction between "unpredictable" and "not predictable", see Gonzalez (2010, p. 289).

[5]On the different predictive processes, see Guillan (2017, ch. 6, 185–215).

[6]In his judgment, there are limits to prediction in social sciences that "lie in the intractability of the issues, so that there is little reason to think that the relatively modest record of the past will be substantially improved upon the future" (1998a, p. 202).

5.2.2 The Social Realm and the Obstacles to Prediction

Although Rescher's main interest regarding prediction is in the realm of natural sciences, he also addresses the problem with regard to the social sciences (above all, economics and sociology) [1998a, pp. 193–202]. The main problems to prediction in this domain are—for him—of an ontological kind, instead of been epistemological problems (1998a, 202), and they have methodological repercussions. Wenceslao J. Gonzalez has pointed out the ontological roots of the methodological problem in the case of economics from the perspective of complexity: "complexity contributes to the frequent lack of reliability of economic predictions, which has its roots in the object of study of this science: economic reality is a social and artificial undertaking, which is commonly mutable, as a consequence of its dependence on the human activity that develops historically" (Gonzalez 2012c, p. 92).

Certainly, when prediction is about the social realm, it has to deal with the complexity of the social systems. In this regard, it is usual to consider that complexity in the social realm is generally greater than the complexity of natural phenomena (cf. Gonzalez 2011a, 2013a). From this perspective, historicity gives us the appropriate framework to address the complexity of the social realm, since it must be considered as a main ontological feature of the social reality. Historicity affects both structural complexity and dynamic complexity (cf. Gonzalez 2012d). This feature has clear repercussions on the possibility of predicting the social systems and on the kind of predictions that we can obtain (with regard to their reliability, accuracy, precision, etc.), as well as on the variety of predictive processes that we use to predict in the social realm (cf. Gonzalez 2012c). According to Rescher, "the difficulties that the predictive project encounters in [natural] science pale in comparison with those it encounters in human affairs" (1998a, p. 192).

When the focus is on the social sciences, historicity sets limits to methodological universalism in the three aforementioned levels: (a) science in general; (b) a group of sciences; and (c) specific sciences. It has been pointed out that historicity is a feature of science, in general, and of each science, in particular. It can be seen in the structure of science (i.e., the configuration of its constitutive elements) and in its dynamics (both from an internal perspective—in terms of aims, processes, and results—and from an external viewpoint—the relations with the context) (Gonzalez 2012b, 7–30; especially, 13–14).

Historicity is also a feature that configures the agents who develop the scientific research, since they are human beings in a concrete historical context. It affects both the approaches to address the research topics and the circumstances that surround the scientific research. Finally, historicity modulates the reality itself that is researched, above all, when it is a social or artificial reality (cf. Gonzalez 2011b, 43). In effect, social reality is historical in itself, so that historicity is a key ontological feature of social systems. This means that social reality has in itself a component of variability that adds methodological and epistemological complexity to prediction in this realm (Guillan 2017, 224).

From this perspective, there are important differences between the objects of natural and social scientific activity. This leads us to dismiss the possibility of methodological universalism. Certainly, the diversity and complexity of the objects and problems of scientific activity are not compatible with a proposal of methodological universalism (characterized as a common method shared by all the empirical sciences) or even methodological imperialism (for example, a dominant method from economics which can be used to address the problems of all the social sciences) [Gonzalez 2020, p. 255].

5.2.3   Prediction in the Artificial Realm

Prediction is also a main aim within the realm of the sciences of the artificial. But, insofar as they are applied sciences, prediction is also a tool for decision-making. In this way, prediction is usually the previous step to prescription.[7] Thus, the anticipation of the possible future (prediction) is required before we give indications about what should be done (prescription) when we want to solve a concrete problem (see Gonzalez 2008, 165–186; especially, 179–183).

A characterization of the artificial realm from an ontological perspective is offered by Herbert A. Simon. He establishes a comparison between artificial things and natural phenomena. For him, there are four main differences between them: "1. Artificial things are synthesized (though not always or usually with full forethought) by human beings. 2. Artificial may imitate appearances in natural things while lacking, in one or many respects, the reality of the latter. 3. Artificial things can be characterized in terms of functions, goals, adaptation. 4. Artificial things are often discussed, particularly when they are being designed, in terms of imperatives as well as descriptives" (Simon 1996, 5).

According to these features, the sciences of the artificial are disciplines that have to do with the reality of the *human-made* (Simon 1996, 4–5. See also Simon 1995). Thus, their field is the realm of what is made by humans. In the same way as it happens with the field of the social sciences, the ontological aspect of historicity is a key feature in the sciences of the artificial. According to Gonzalez, "insofar as it is a realm of specific human elaboration that seeks to enlarge the existing potentialities or to achieve new and more ambitious aims, there is always a contextual component that is modulated by historicity" (2012b, 14). However, Simon did not manage to grasp this feature of historicity, which goes with the creativity of the designs.

A specific feature of the sciences of the artificial is that they are open to novelty, insofar as those sciences (especially, when they are configured as sciences of design) are disciplines that actively search for new possibilities to enlarge the reality (they are not limited to achieve knowledge of the researched reality). In effect, the reality of the artificial realm is changeable, and novelty can appear through the creativity

---

[7]See, in this regard, Niiniluoto (1993, 1995).

of the agents (see Guillan 2013, 125–139). Although Rescher does not address the field of the sciences of the artificial, his approach to scientific prediction takes into account creativity as an obstacle to prediction. For him, "human creativity and inventiveness defies predictive foresight" (Rescher 1998a, 149).

Ontologically, historicity and creativity are two main characteristics of the artificial realm, and they are directly related to the possibility of novelty in the artificial field. These features add complexity to the task of predicting in the sciences of the artificial. The epistemological and methodological repercussions are clear: firstly, the continuous presence of novelty and changes related to creativity and historicity affect the achievable predictive knowledge (its reliability, accuracy, precision, etc.); and, secondly, predictive procedures and methods must take into account the ontological features of the reality that is researched, where creativity, historicity, and the relations with the context have an important role (Guillan 2017, p. 225).

Thus, when the focus is on the reality of the phenomena (natural, social, and artificial), it seems clear that there are limits to methodological universalism in the three levels that have been pointed out: science in general, a group of sciences (natural sciences, social sciences or sciences of the artificial), and each concrete science (biology, economics, pharmacology, etc.). This is because the choice of the predictive procedures and methods must take into account the characteristic features of the reality that we want to predict. This involves methodological variations, which pose difficulties to methodological universalism in all the mentioned levels.

Methodologically, Rescher has developed an approach of methodological pragmatism (1977), which is not compatible with an account of methodological universalism (i.e., the possibility of having a method with general validity for all the empirical sciences). Regarding scientific prediction, his methodological proposal goes in two directions: on the one hand, he seeks to clarify the common features of the different predictive procedures and methods; and, on the other hand, he wants to make explicit the distinctive features of each concrete predictive process (see Guillan 2017, Chaps. 5 and 6). Within the second direction, he makes a distinction between estimative procedures of prediction and discursive or formalized predictive processes, which can be elementary processes or scientific methods (Rescher 1998a, pp. 86–88). This is an approach that assumes, *de facto*, a methodological pluralism regarding scientific prediction.

To sum up, when the different realms of reality are considered, there are limits to methodological universalism with regard to science in general, a group of sciences, and each science in particular. Firstly, there is variety in the predictive processes in accordance with the studied realm, because the natural, social, and artificial phenomena have different characteristics. From this perspective, it seems difficult that we can have dominant processes with general validity for the whole set of the empirical sciences. Secondly, within each group of sciences (natural sciences, social sciences or sciences of the artificial), the different sciences can address phenomena with different features. Thus, for instance, phenomena that we want to predict in astronomy are generally more stable than phenomena studied by meteorology. This affects the methods oriented to predict in the different disciplines within a concrete realm (natural, social or artificial).

Thirdly, when the analysis is directed towards a concrete science, it can be seen that variations and obstacles to predictability can also appear. They can be both epistemological and ontological, and certainly they have methodological repercussions. On the one hand, the temporal range of the prediction should be considered. Prediction can be, in principle, in the short, medium or long run. And, on the other hand, predictive models can vary in accordance with the different levels of reality (micro, meso, and macro). These considerations lead to the problem of complexity, which is especially relevant to address the limits of the methodological universalism.

5.3 Ontology of Prediction as a Support for the Methodology of Science: The Perspective of Complexity

Certainly, the different proposals of methodological universalism have to consider the specific features of the realms of reality (natural, social, and artificial), as well as the ontological variations within each one of those realms. Furthermore, when the focus is on scientific prediction, they have to take into account the differences between basic science, which seeks predictions in order to increase the available knowledge, and applied science, where prediction is usually the previous step to prescription (cf. Gonzalez 2013b). From this perspective, there are limits to methodological universalism regarding scientific prediction, which can be analyzed from the angle of complexity (ontological, epistemological, and methodological).

Prediction can be about a reality (natural, social or artificial) that can be characterized as complex, so its complexity has repercussions that goes in two directions: on the one hand, it has repercussions on the very possibility of predicting; and, on the other, it modulates the configuration of the predictive processes and the type of prediction that can be achieved (its reliability, accuracy, precision, etc.). In this regard, it should be taken into account that complexity is a two-fold notion: it has a structural dimension (the components of the complex system and their interactions) and a dynamic trait (the changes that occur in the system over time, both from an internal point of view and from an external perspective) (cf. Gonzalez 2013a).

5.3.1 Varieties and Relevant Modes of Complexity

According to W. J. Gonzalez, "complexity is indeed a very important source of difficulties for methodological universalism because it affects problems, methods and results of the scientific research" (2012a, 162). When the complexity of a system (natural, social or artificial) is considered, two different dimensions of complexity should be taken into account: (a) the structural dimension (that has to do with the parts that compound the system and their interactions); and (b) the dynamic trait (that deals with the change over time in the system).

Concerning the structural dimension, Rescher highlights that there are numerous features of complexity. He distinguishes two basic levels with regard to the modes of complexity: the epistemic and the ontological ones. Epistemic complexity— in Rescher's account—is related to formulas. His proposal offers three epistemic modes of complexity: (i) descriptive, (ii) generative, and (iii) computational. Descriptive complexity is related to the level of difficulty that involves describing a system. Generative complexity deals with the number of steps that are required to give rise to the complex system, and computational complexity has to do with the number of resources that are needed in order to solve problems related to the system (Rescher 1998b, 9).

Regarding the ontological aspect, Rescher also distinguishes three ontological modes of complexity: (a) compositional complexity, which is related to the elements that compound the system and their variety; (b) structural complexity, which deals with the organization and relation between the elements of the complex system; and (c) functional complexity, which is related to the behaviour of the system (Rescher 1998b, 9). In turn, each one of them has two possible options. Thus, compositional complexity can be constitutional (related to the number of elements that compound the system) or taxonomic (associated with the variety of those elements) (1998b, 11). Within structural complexity, Rescher distinguishes organizational complexity (related to the forms of organization that can be seen in the system), and hierarchical complexity (which has to do with the ways of relating the components of the system) (1998b, 11–12). Finally, functional complexity can be operational complexity and nomic complexity. Operational complexity studies the variety of the modes of operation and the types of functioning that appear in the system, an issue that Rescher sees in terms of processes. Nomic complexity is related to the patterns that regulate the relations between the elements of the system (Rescher 1998b, 12–13).

Rescher's account of ontological complexity highlights that, when we try to encompass the complex reality of a complex system (natural, social or artificial) there are a wide variety of factors to take into account. Thus, the ontological complexity of a system involves that we have to consider the components (their number and variety), the structure (regarding the organization and the relations between the components), and the functions (the types of operation and the working patterns).

Ontological complexity affects prediction, since prediction of a complex system with ontological complexity has to deal with a wide variety of factors (regarding the composition, the structure, and the function of the system). This is because, although a system does not need to be complex in all the aforementioned senses, "the different modes of complexity do tend to run together in practice" (Rescher 1998b, 15). This means that, when a system is complex with regard to its composition and structure, it is usually complex regarding their function too (Rescher 1998b, 15). Certainly, this makes it difficult to predict, and has repercussions on issues such as the reliability or the degree of accuracy and precision that can be achievable and on the predictive processes (Guillan 2017, pp. 242–243).

The *methodological modes* of complexity can be added to Rescher's proposal, since the complexity related to reality and knowledge has repercussions on the scientific processes (for example, in the case of the predictive models). Methodological universalism has to consider that methodology of science deals with complex issues. As it has been pointed out by Gonzalez, "each mode of complexity—epistemic or ontological— can pose some difficulty for the universal methodology, regarding generality and reliability. The methodology of science needs to deal with issues that are not simple, which might be at different epistemological levels and can belong to diverse stages of reality. The researcher uses processes that depend on the objects (the aspect of reality) and the kind of problem (the focus of attention)" (Gonzalez 2012a, 164).

5.3.2 The Dynamic Trait of Complexity

There are still more factors to consider than only the epistemic, ontological, and methodological modes of complexity. Certainly, together with the structural dimension of complexity, there is a dynamic trait. In turn, when dynamic complexity is considered, we must take into account, in principle, two levels: the "internal" dynamics and the "external" dynamics (cf. Gonzalez 2012b, 13–14). The internal dynamics is related to the activity of the system itself (mainly in terms of aims, processes, and results), whereas the external dynamics deals with the relations of the system with the context.

The concept of "historicity" has been proposed to characterize the change in the complex dynamics (both "internal" and "external") and its repercussions on scientific prediction (Gonzalez 2012c, 79–88). It is a notion that goes further than the mere temporality, since it is difficult to think of a reality that escapes from the temporal dimension. However, historicity is a characteristic of the human realm: it links with the human activity, which is historic, in the sense that it changes over time (Gonzalez 2003, 88). In addition, historicity has into account the context that surrounds human activity.

This notion is related to three levels of analysis: (a) science, (b) the agents, and (c) the reality itself that is researched. Gonzalez has pointed out that "(1) Historicity (*Geschichtlichkeit/historicidad*) is a trait of science, in general, and each science, in particular. This facet can be found in the whole set of constitutive elements of science, such as language, structure, knowledge, method, activity, ends, and values. (2) Historicity configures the agents themselves involved in the development of scientific research, insofar as they are human beings within a historical context. (3) Historicity is a characteristic of the reality itself that is researched (above all, in the social and artificial realms)" (Gonzalez 2011b, 43).

From this perspective, it seems clear that the notion of *historicity* is relevant when both the problem of complexity and its repercussions on scientific prediction are considered. As it has been pointed out above, historicity is a feature that modulates science. This feature can be seen in three successive levels: the constitutive elements of science (that are related to the structural dimension of complexity), the internal dynamics of the scientific activity (in terms of aims, processes, and results), and the external dynamics (the relations with the changing context).

Additionally, historicity configures the agents who develop the scientific activity, who are in a concrete socio-historical context, which is changeable. This feature can add more complexity (for instance, through human creativity, both individual and social, and novelty). Finally, historicity can be a feature of the reality itself studied by science, above all when it is a social or artificial reality. From this perspective, historicity can be the basis to understand why prediction is more difficult (and generally less reliable) in social sciences and in the sciences of the artificial, where the complex systems that we want to predict are historical (Guillan 2017, 244–245).

Furthermore, the historical dimension also sets limits to methodological universalism. In this regard, we can distinguish a methodological universalism with a timeless component and a methodological universalism within the context of historicity (cf. Gonzalez 2012a, 159–162). In the first case, the historical dimension of the methodology of science is completely dismissed; whereas in the second case the methodological universalism is expressed in historical terms. Moreover, "the forms of methodological universalism in science can change historically" (Gonzalez 2012a, 161), so what is *de facto* a dominant methodological approach can be finally replaced by another methodological perspective.

5.3.3 Characteristics of Complexity in the Social Sciences and in the Sciences of the Artificial: The Case of Communication Sciences

Complexity is a key issue for the study of the social sciences and the sciences of the artificial. In fact, these sciences can be analyzed as sciences of complexity, both with respect to their structure and with regard to their dynamics (internal and external). On the one hand, they are sciences with structural complexity. This can be seen in relation to the constituents of a science (language, internal articulation, knowledge, method, activity, ends, and values) (Gonzalez 2012c, 74). On the other hand, their dynamics (both internal and external) is complex, above all when those sciences are developed as applied sciences (cf. Gonzalez 2012b, 8–9). Thus, in their internal dynamics, there is an articulation of aims and processes, which generate a result. At the same time, they involve an external dynamics, since their activity is developed in a changing context.

Both the structural dimension and the dynamic trait of complexity in social sciences and the sciences of the artificial are modulated by historicity. As a human activity, this feature is shared by the natural sciences, since scientific activity is *eo ipso* historic. However, historicity is also an ontological feature of the social and artificial reality. In addition, there is a contextual component. "Because the agents—individual and social—act usually with regard to contexts (historical, cultural, political, . . .), which are changing" (Gonzalez 2012b, 14). From this perspective, historicity allows us to understand the huge variability of the reality (above all, social and artificial), insofar as it offers the appropriate framework to address factors like human creativity or human choices, which are a source of complexity for predictions in social sciences and sciences of the artificial.

However, the studies on complexity, both in social sciences and in the sciences of the artificial, usually do not take into account historicity. Whereas, they are focused in notions such as "evolution" or "process," which do not achieve the depth that historicity achieves (cf. Gonzalez 2012c, 79–84). A more sophisticated analysis of complexity (both structural and dynamic) should be developed through historicity, above all in the social and artificial realms. This can be clearly seen in the case of sciences such as economics or communication sciences, which are dual sciences (they are social sciences and also sciences of the artificial), so they involve features of both groups of disciplines.

Certainly, the dual nature of communication sciences as social sciences and sciences of the artificial involves a source of complexity (Gonzalez and Arrojo 2015, pp. 299–301). On the one hand, communication sciences analyze communication as a human need developed in a social context; and, on the other hand, they enlarge the human abilities of communication. In this regard, they usually need a technological support. This structural complexity can be seen in the configuration of communicative designs, "which combine scientific with technological knowledge for creativity and problem solving. In addition, it has a marked projection due to the social component" (Arrojo 2017, p. 431).

There are also features of complexity in communication sciences due to their configuration as applied sciences of design (Gonzalez and Arrojo 2015, 299–301). As design sciences, communication sciences make new communicative designs (through aims, processes, and results) that create new possibilities; and, as applied sciences, they seek to solve concrete problems of communication. In this regard, there is a complex dynamics, both from an internal perspective and an external viewpoint: "complex dynamics related to the nexus between means and ends in communication sciences cannot be completely isolated from the context (at least legal, organizational, and social)" (Gonzalez and Arrojo 2015, 309).

This issue is related with the distinction between "communicative activity" and "communication as activity," which adds more features of complexity to communication sciences and communicative designs (Gonzalez and Arrojo 2015, 301). On the one hand, *communicative activity* can be understood, in principle, as autonomous regarding other human activities, since it has its own characteristics. On the other hand, *communication as activity* is related with other human activities (cultural, political, social, ecological, etc.). From this perspective, there are relations with the context, so communication as human activity—which is interrelated with other activities— is not independent of the context.[8] Through the "communicative activity," designs have to deal with the modes of complexity that come from the realm of communication. At the same time, "communication as activity" involves that communicative designs have to consider other modes of complexity (social, political, legal, etc.).

In both cases, historicity has a key role, since the communicative reality cannot be understood without the historicity of the human activity made in a social context. Certainly, both the structural dimension and the dynamic trait are modulated by historicity. Communication sciences are the result of human activity, which is *eo ipso* historical. But historicity is also an ontological feature of the reality itself studied by communication sciences, since as dual sciences they study both the social reality and the human made realm (Arrojo 2020).

All these considerations allow us to see how the analysis of complexity from the historicity of the human activity can shed light on the problem of scientific prediction, in general, and prediction in the social sciences and the sciences of artificial, in particular. Certainly, Rescher's approach to the epistemic and ontological modes of complexity allows us to understand some of the methodological difficulties to prediction. But this approach should be incorporated into a broader account of complexity. Thus, we have to consider two aspects of complexity: the structural dimension and the dynamic trait (internal and external). To do this, historicity— in my judgment—should be acknowledged, since it is a key feature of science, the agents, and the reality itself that is researched (above all, social and artificial). All of this reinforces the idea of the existence of limits to the methodological universalism.

5.4 Coda

Methodologically, scientific prediction can be addressed from three different levels of analysis (cf. Gonzalez 2012a, 158): (i) science in general, (ii) a group of sciences; and (iii) a concrete science. When the attention goes to the ontological features of the reality itself (natural, social, or artificial), it seems difficult to assume an explicit methodological universalism (the acceptance of a method as valid for science in general or for a complete domain).[9] In effect, there are different predictive methods and procedures in accordance with the characteristics of the predicted reality (natural, social or artificial), so there are no dominant processes with general validity for the whole set of the empirical sciences.

When the focus is on each group of sciences (natural sciences, social sciences, and the sciences of the artificial), the studied phenomena have also different features. These features affect the kind of processes to be used, and the expected results of

---

[8]See also Arrojo (2012). This distinction is related with the distinction made in economics between "economic activity" and "economics as activity". See Gonzalez (1994).

[9]On the different varieties of methodological universalism see Gonzalez (2012a).

the predictive activity. Complexity—both structural and dynamic—also sets limits to proposals about a dominant method within each group of sciences or a concrete science (cf. Gonzalez 2012a, 2013b), as can be seen when scientific prediction is considered.

The relevance of the perspective of complexity can be clearly seen in the concrete case of the communication sciences. From the structural perspective, communicative phenomena are ontologically dual: artificial (due to the designs), and social (due to their public impact) (cf. Arrojo 2017; Gonzalez and Arrojo 2019). Thus, communication sciences are dual sciences: they are both social sciences and sciences of the artificial (cf. Gonzalez and Arrojo 2015). In this regard, the epistemological, ontological, and methodological modes of complexity affect communication sciences and set limits to prediction (cf. Gonzalez 2013b).

Furthermore, they are applied sciences of design, as they develop communicative designs that open new possibilities and seek to solve concrete problems of communication. From this perspective, prediction is also required as a previous step to prescription. In this regard, there is a complex dynamics in communication sciences due to the interaction between aims, processes, and results (internal dynamics) and also due to the relations with the legal, organizational, and social context (external dynamics) (cf. Gonzalez and Arrojo 2015, 2019). Here, historicity (as a feature of science, the agents, and the reality itself that is researched) sheds light on the problem of complexity and, consequently, on the analysis of scientific prediction and the methodological difficulties.

References

Arrojo, M. J. (2012). La sobriedad de factores en el análisis de la complejidad en las Ciencias de la Comunicación. El estudio de la Televisión Digital Terrestre. In W. J. Gonzalez (Ed.), *Las Ciencias de la Complejidad: Vertiente dinámica de las Ciencias de Diseño y sobriedad de factores* (pp. 337–358). A Coruña: Netbiblo.

Arrojo, M. (2017). Information and the internet: An analysis from the perspective of the science of the artificial. *Minds and Machines, 27*, 425–448.

Arrojo, M. (2020). Objectivity and truth in sciences of communication and the case of the internet. In W. J. Gonzalez (Ed.), *New approaches to scientific realism* (pp. 415–435). Berlin/Boston: de Gruyter.

Gonzalez, W. J. (1994). Economic prediction and human activity. An analysis of prediction in economics from Aqction theory. *Epistemologia, 17*(2), 253–294.

Gonzalez, W. J. (2003). Racionalidad y Economía: De la racionalidad de la Economía como Ciencia a la racionalidad de los agentes económicos. In W. J. Gonzalez (Ed.), *Racionalidad, historicidad y predicción en Herbert A. Simon* (pp. 65–96). A Coruña: Netbiblo.

Gonzalez, W. J. (2008). Rationality and prediction in the sciences of the artificial. In M. C. Galavotti, R. Scazzieri, & P. Suppes (Eds.), *Reasoning, rationality and probability* (pp. 165–186). Stanford: CSLI Publications.

Gonzalez, W. J. (2010). *La predicción científica. Concepciones filosófico-metodológicas desde H. Reichenbach a N. Rescher*. Barcelona: Montesinos.

Gonzalez, W. J. (2011a). Complexity in economics and prediction: The role of parsimonious factors. In D. Dieks, W. J. Gonzalez, S. Hartman, T. Uebel, & M. Weber (Eds.), *Explanation, prediction, and confirmation* (pp. 319–330). Dordrecht: Springer.

Gonzalez, W. J. (2011b). Conceptual changes and scientific diversity: The role of historicity. In W. J. Gonzalez (Ed.), *Conceptual revolutions: From cognitive science to medicine* (pp. 39–62). A Coruña: Netbiblo.

Gonzalez, W. J. (2012a). Methodological universalism in science and its limits. Imperialism versus complexity. In K. Brzechczyn & K. Paprzycka (Eds.), *Thinking about provincialism in thinking, Poznan studies in the philosophy of the sciences and the humanities* (Vol. 100, pp. 155–175). Amsterdam/New York: Rodopi.

Gonzalez, W. J. (2012b). Las Ciencias de Diseño en cuanto Ciencias de la Complejidad: Análisis de la Economía, Documentación y Comunicación. In W. J. Gonzalez (Ed.), *Las Ciencias de la Complejidad: Vertiente dinámica de las Ciencias de Diseño y sobriedad de factores* (pp. 7–30). A Coruña: Netbiblo.

Gonzalez, W. J. (2012c). La vertiente dinámica de las Ciencias de la Complejidad. Repercusión de la historicidad para la predicción científica en las Ciencias Diseño. In W. J. Gonzalez (Ed.), *Las Ciencias de la Complejidad: Vertiente dinámica de las Ciencias de Diseño y sobriedad de factores* (pp. 73–106). A Coruña: Netbiblo.

Gonzalez, W. J. (2012d). Complejidad estructural en Ciencias de Diseño y su incidencia en la predicción científica: El papel de la sobriedad de factores (*Parsimonious Factors*). In W. J. Gonzalez (Ed.), *Las Ciencias de la Complejidad: Vertiente dinámica de las Ciencias de Diseño y sobriedad de factores* (pp. 143–167). A Coruña: Netbiblo.

Gonzalez, W. J. (2013a). The sciences of design as sciences of complexity: The dynamic trait. In H. Andersen, D. Dieks, W. J. Gonzalez, T. Uebel, & G. Wheeler (Eds.), *New challenges to philosophy of science* (pp. 299–311). Dordrecht: Springer.

Gonzalez, W. J. (2013b). Los límites del universalismo metodológico: El problema de la complejidad. *Naturaleza y Libertad. Revista de estudios interdisciplinares, 2*, 61–89.

Gonzalez, W. J. (2015). *Philosophico-methodological analysis of prediction and its role in economics.* Dordrecht: Springer.

Gonzalez, W. J. (2020). Pragmatic realism and scientific prediction: The role of complexity. In W. J. Gonzalez (Ed.), *New approaches to scientific realism* (pp. 251–287). Berlin/Boston: de Gruyter.

Gonzalez, W. J., & Arrojo, M. J. (2015). Diversity in complexity in communication sciences: Epistemological and ontological analyses. In D. Generali (Ed.), *Le radici della razionalita critica: Saperi, Pratiche, Teleology* (pp. 297–312). Milan-Udine: Mimesis Edizioni.

Gonzalez, W. J., & Arrojo, M. J. (2019). Complexity in the sciences of the internet and its relation to communication sciences. *Empedocles: European Journal for the Philosophy of Communication, 10*(1), 15–33.

Guillan, A. (2013). Analysis of creativity in the sciences of design. In W. J. Gonzalez (Ed.), *Creativity, innovation, and complexity in science* (pp. 125–139). A Coruña: Netbiblo.

Guillan, A. (2017). *Pragmatic idealism and scientific prediction: A philosophical system and its approach to prediction in science.* Dordrecht: Springer.

Niiniluoto, I. (1993). The aim and structure of applied research. *Erkenntnis, 38*(1), 1–21.

Niiniluoto, I. (1995). Approximation in applied science. *Poznan Studies in the Philosophy of Sciences and the Humanities, 42*, 127–139.

Rescher, N. (1977). *Methodological pragmatism. A systems-theoretic approach to the theory of knowledge.* Oxford: Blackwell.

Rescher, N. (1978). *Scientific progress. A philosophical essay on the economics of the natural science.* Oxford: Blackwell.

Rescher, N. (1998a). *Predicting the future. An introduction to the theory of forecasting.* New York: State University of New York Press.

Rescher, N. (1998b). *Complexity: A philosophical overview.* N. Brunswick: Transaction Publishers.

Rescher, N. (1999a). *Razón y valores en la Era científico-tecnológica.* Barcelona: Paidós.

Rescher, N. (1999b). *The limits of science* (rev. ed.). Pittsburgh: University of Pittsburgh Press.

Rescher, N. (2006). Pragmatic idealism and metaphysical realism. In J. R. Shook & J. Margolis (Eds.), *A companion to pragmatism* (pp. 386–397). Oxford: Blackwell.

Rescher, N. (2009). *Unknowability. An inquiry into the limits of knowledge*. Lanham: Lexington
    Books.
Rescher, N. (2010). *Reality and its appearance*. London: Continuum.
Simon, H. A. (1995). Problem forming, problem finding, and problem solving in design. In A.
    Collen & W. W. Gasparski (Eds.), *Design and systems: General applications of methodology*
    (Vol. v. 3, pp. 245–257). New Brunswick: Transaction Publishers.
Simon, H. A. (1996). *The sciences of the artificial* (3rd ed.). Cambridge, MA: The MIT Press. (1st
    ed., 1969; 2nd ed., 1981).

Part III
Contextual Factors for Methodology of Science

Part III

Chapter 6
Information and Pluralism. Consequences for Scientific Representation and Methods

Giovanni Camardi

Abstract Both logic and philosophy of science have developed independent pluralistic views. Nevertheless, a general notion of pluralism is quite elusive. This paper will draw on the theory of information, including Shannon's version and its fundamental concept of *coding*, to set up a notion of pluralism that may contribute to an updated idea of the scientific method, such as to interact with computational models. I will discuss Patrick Allo's notion of "informational pluralism" and connect it with Wesley Salmon's treatment of "statistical relevance" and James Woodward's analysis of the "data-phenomena" relationship. Based on the concept of information, I will argue that there is a possible convergence of the three views.

Keywords Pluralism · Information · Logic · Coding · Models · Relevance

6.1 Introduction: Logical and Scientific Pluralism

Logic and philosophy of science have independently developed pluralistic views. Logicians have reacted to the monism of classical logic, while philosophers have given up the ideal of a unique scientific method. But these moves could have undesirable consequences, namely inflation of logical objects and a shallow *de facto* pluralism in scientific research.

Yet, there could be room for discovering synergies.

Current studies on logical pluralism might be considered an attempt to provide new rules and limit post-classical fragmentation. Logical pluralism may improve on the *de facto* pluralism in science. Furthermore, *informational pluralism* – a combination of logical pluralism and information theory, advocated by Patrick Allo and Edwin Mares – could be the right framework for understanding the role of models, which have become a bridge between logic and science. Models, mainly because they come in computational clothing, are currently predominant

G. Camardi (✉)
Department of Humanities, University of Catania, Catania, Italy

© Springer Nature Switzerland AG 2020   99
W. J. Gonzalez (ed.), *Methodological Prospects for Scientific Research*, Synthese
Library 430, https://doi.org/10.1007/978-3-030-52500-2_6

over theories (Morrison 2015). Hence, the notion of "representation" has proved to be central both in logic and information (van Fraassen 2008). Ultimately, scrutiny of logical and informational pluralism could be an appropriate framework to assess the impact of information theory on scientific practice.

However, the notion of pluralism is not as intuitive as it might seem at first glance. It is, instead, quite elusive (Caret 2019). There is no canonical account of a pluralist system (Beall and Restall 2000), and there are varieties of pluralism possibly independent from one another (Cook 2019). We may even say that a semantically complete definition of "pluralism," if any, would be at risk of being inconsistent. On the other extreme, accepting an unconstrained pluralism would end up with the elimination of any form of logical rigor, and as a result, pluralism may collapse in some sort of "logical nihilism" (Russell 2019). Thus, we have to find a way to navigate this slippery terrain.

6.2   Logical Consequence, Models, Information

Logical pluralism consists in assuming that there is more that one logical consequence relation and in determining how to operate each of them across arguments, computations, and communications. The scheme of logical consequence (LC) is expressed in a principle, called "General Tarski Thesis" (GTT): "An argument is valid, if and only if, in every case$_x$, in which the premises are true, so is the conclusion." We can infer that logical pluralism consists of "the claim that at least two different instances of GTT provide admissible precisifications of logical consequence." (Beall and Restall 2006, 29) This is the basis of logical pluralism.[1] Our job is now to determine what kinds of *cases* can preserve logical consequence, i.e., what the subscript "$_x$" means. More explicitly, we have to spell out the meaning of "in every case$_x$" in GTT and put constraints on it. What Beall and Restall call "precisification" consists of specifying "the cases$_x$ over which GTT quantifies" and what it means for a claim to be true in that sort of cases. Beall and Restall suggest that "cases" may be either "possible worlds" or "set-theoretic constructions, models of some sort." (Beall and Restall 2006, 36).

I will go for the second possibility and take "cases" as models. Indeed, the connection between informational pluralism and scientific representation I intend to discuss in the second part of this paper needs a model-theoretic approach. A pure logic of possible worlds is not enough to explore the structural affinities between informational pluralism and the visionary project in scientific methodology

---

[1]Beall and Restall's formal account of logical pluralism provides the best opportunities for an information-theoretic approach, directed to support scientific methodology. Consequently, this paper will focus on semantic and model-theoretic issues and will be less concerned with modal topics such as necessity, possibility, and consistency. In this line of thought, modal logic will be reduced to a few mentions, mostly concerning the accessibility relation. For an alternative view and a more intense connection between pluralism and modal logic, see Bueno and Shalkowski (2009).

pursued by Wesley Salmon as early as 1984. Salmon had based his statistical-relevance or S-R model of scientific explanation on an analysis of statistical data, and his concept of causation on a heuristic analogy between causal propagation and transmission of information. A few years later, James Woodward's "Data and Phenomena Reasoning" was developed along the same lines as Salmon's.

Informational pluralism as well has been developed in terms of *relevance logic*, which considers models and cases as *situations*. A model contains "sets of situations" while situations contain "information as represented by the formulae of a formal language." Hence, objective information consists of "potential data available in an environment." (Mares 2010, 113).[2]

Nevertheless, we need to learn more about the relevance of situations. Besides the elucidation of *cases*, Gillian Russell has provided a more fine-grained analysis of logical consequence. She has discerned four accounts of LC, depending on how the non-logical expressions of an argument are defined.[3] Each of them is not alternative to, or exclusive of the others. We can use them at various moments of the development of a model. A choice depends on personal interests. We are now using Russell's interpretational and representational accounts (Russell 2019) because we are interested in making sense of the link between logical pluralism and scientific representations, while a model is under construction. So, we are defining the properties of the types, picking up those appearing more helpful in configuring data to be entered. Assuming logical pluralism as a premise, we may use different schemes of LC to select and combine various *interpretations* of non-logical predicates (properties). Then we may use LC to *represent*, i.e. encode raw data and subsequently recode them as information data to be fitted into the types of a computational model. Thus, Russell's interpretational and representational accounts have made more readable to us the two crucial stages of model construction.

Still, Russell's differentiation of the two accounts reminds us that models present us with a challenge. There is a discontinuity between the interpretative and representative levels, and we have to fit types into data and vice versa, smoothing out the incongruous traits present in both of them. Computational models have made us familiar with this job. When we deal with them, there may be several elements of non-linearity, and we have to account for them (see p. 6, below). Now, data to be entered in a model are not "raw" data. Perhaps we better call them *factual* data, *represented* and encoded, at a basic level. Most appropriately, Mares points out that factual data are "environmentally available information," a sort of "bottom-up

---

[2]If objective information consists of potential data, they plausibly become actual when *encoded* in the formulae of a language. See Sect. 6.4, below.

[3]Russell distinguishes (1) a *substitutional* account of LC, where non-logical expressions are uniformly substituted throughout the argument by "syntactically appropriate alternative expressions"; (2) an *interpretational* account, where different meanings replace the meanings of certain non-logical expressions; (3) a *representational* account, where non-logical expressions are represented by various situations or states of the world; (4) a *universal* account, where the non-logical vocabulary is uniformly substituted by different assignments of values to non-logical variables. See Russell (2019), 337–339.

justification" for relevant relations of logical consequence (Mares 2010, 117). Thus, LC is constrained by an "accessibility" relation that is shown by the "$_r$" subscript, $\Phi \rightarrow_r \Psi$, and represents a situated inference (Mares 2010, 125). At this point, computational modeling and multiple programming languages have shifted away command, from logicians towards modelers. But we have the necessary premises to introduce informational pluralism. We are learning that computational models are the frames where the intervention of information in scientific practice is designed to happen. Computational models will undoubtedly support an informational and pluralist analysis of scientific practice. They are a framework for holding together the logically-based construction of computable types and systems on the one hand, and the statistical and computational collection of data, on the other (see Turner 2009; O' Donnell 1998; Cover and Thomas 2006, chaps. 5 and 11).

To sum up, Allo and Mares' informational pluralism and Salmon's statistical relevance approach share two fundamental premises – relevance logic, and information theory. Also, informational pluralism is designed to be expressed in model-theoretic structures, like Salmon's scheme of propagation of causal influence (Salmon 1977), paralleled by the process of transmission of information, and satisfied in a model integrating relevant statistical data into computable types. I submit that Allo and Mares' studies on informational content may complete the informational segment of Salmon's information-causation speculative parallel and, thereby, the *de facto* pluralism in scientific modeling.

I claim, however, there is a critical condition for this process to be successful: the conception of information that Allo, Mares, Restall and Barwise (among others) have adopted, is to be updated, under the penalty of failing to address the passage from logical to informational pluralism. Their logical analysis of information – however correct it may be – may mistake information for a sort of logical object. I think this is unsound. Barwise has appropriately recalled that "information is not a syntactic property of sentences" (Barwise and Seligman 1997, 7). There is a feature of *information* that has to be highlighted, and this is the concept of *code*. A central character of transmission of information is that messages must necessarily be encoded. Claude Shannon, the founder of the theory, has given encoding a central place, by linking "proper encoding" to "channel capacity."

I will now proceed this way. First, I will discuss in more detail Allo's concept of informational content. Then I will address the possibility that the notion of code might be fastened to specific enhanced versions of that concept proposed by Barwise, Mares, and Allo himself.

6.3   Informational Content and Informational Pluralism

Allo's first point is the dismissal of the monism of classical logic, namely the presumption that there is a unique formalization of logical consequence so that no constraint on it is required. Consequently, Allo accepts Beall and Restall's version of logical pluralism and plans to take it further. To support his concept of informational pluralism, he sets up a logical structure that employs the *relevance logic* as an alternative to classical logic. If the latter is monistic, the former is, indeed, pluralistic. In classical logic, the set of consequences A that can be derived from a set of premises Σ is complete and consistent. But in relevance logic, consequences are assumed to be incomplete and possibly inconsistent. State-descriptions or "cases" are neither complete and consistent possible-worlds nor intuitionistic constructions. Cases are incomplete factual situations (Allo 2007, 660; Beall and Restall 2006). Thus, informational content[4] – which consists of well-formed, meaningful and truthful data (Allo 2007, 661) – sits in situations, considered as "partial information states" (Allo 2007, 665), requiring a logical evaluation (⊩) to distinguish meaning and information.[5] Now, the "information" to be evaluated is a property of all kinds of data. Shannon called it "entropy," and it consists of a measurable capacity of producing a reduction of uncertainty associated with the possible outcomes of a sequence of events. Intuitively, an increase in information produces a decrease in possible outcomes. Hence, there is an inverse relationship between available information and possible outcomes. As a result, informational content can be synthesized in a principle that can be traced back to Shannon, Barwise or even Popper, the *Generalized Inverse Relationship Principle* (GIRP): "The informational content of a piece of information, is given by the set of cases it excludes." (Allo 2007, 662).

We now have a sufficient understanding of the concept of *informational content* and of its relationship with the more elusive and controversial notion of *information* from which it originates. I will not track the latter and focus instead on informational pluralism.

For informational pluralism to work correctly, the logical structure that supports it ought to be in control on the one hand of the partiality and incompleteness of information states and, on the other, of the uncertainty related to data collection and processing.

Communication between partial information states is an essential feature of both information theory and relevance logic and, therefore, it requires a strong commitment to investigate the inherently incomplete and fragmentary notion of situated informational content. If logical pluralism is to be maintained, the logical consequence is to remain valid, content is not to be expanded and the truth is to be preserved across a network of partial situations, then Allo has to disambiguate several intricacies and reconcile the ensuing dichotomies to the purpose of keeping informational pluralism credibly in place. In addition, it would be helpful to imple-

---

[4]The concept of informational content has been used by Dretske in his classical analysis. He distinguishes between meaning and information and defines information as "a commodity capable of yielding knowledge" (Dretske 1981, 44). Allo accepts this distinction (Allo 2007, 687), which involves the problem of defining "information." I am not going to pursue such a thorny question in this paper.

[5]Allo shares Floridi's conception of information as "declarative, objective and semantic information" (See Floridi 2011). But he is also well aware of Dretske's classical analysis (cf. Dretske 1981).

ment more versatile semantic structures to make room for the intrinsically variable data that may result from significantly complex statistical modeling. I am going to discuss first the versatility issue and afterward the question of communication and semantic stability.

In order to achieve a flexible structure, Allo uses an intensional operator, *, established by Routley and Meyer, and he sets up a *Star Information Frame*:

An information frame **F** is a structure $(S, Log, \sqsubseteq, *)$ where S is a non-empty set of situations, Log is the subset of logical situations, $\sqsubseteq$ is an ordering relation on S such that if $s_1 \sqsubseteq s_2$ then s_2 contains at least as much information as s_1, and * is a unary operator on S such that:

$$s_1 \sqsubseteq s_2 \text{ iff } s_2^* \sqsubseteq s_1^*$$

Informally, s^* is a sort of counterpart of s, which can be understood as asserting all that s does not deny. Hence, s_1 is an informational order on s_2, i.e. s_2 contains as much information as s_1 if and only if all that s_1 does not deny contains as much information as all that s_2 does not deny. Thus, the * operator "encodes what it means for two situations s_1 and s_2 to be compatible" (Allo 2007, 667). The payoff of this procedure consists in extending the informational limits of a frame F, adding up to the set of situations included in it a set of situations logically compatible with them. We can cash in fresh informational data derived from entailment relations provided by relevance logic, data we could not have cashed through classical logic. In so doing, Allo strikes a balance between expanding to the maximum the informational content of a *situation* and preserving its logical cogency, based on truth-preservation or content-non-expansion, which "ultimately come down to the same thing." (Allo 2007, 687).

Let us now consider the problem of the fragmentary nature of relevance logic. The solution is based on two factors. The first is the endurance of the "worldly" perspective of classical logic—which cannot be repudiated and remains the unshaken ground of every logical innovation and is, therefore, entirely consistent with the logical pluralism we are defending. The second factor is a concept that was conceived in the field of information theory and then adopted by several logicians, under a different form. This is the concept of a *channel*. Allo has come across it in the course of his search for a common language that might allow optimal communication between different kinds of logics. His analysis is first run in a semantic key and built on the contrast between the requirement of a context-independent "fixed meaning" that is supposed to secure the stability of communication on the one hand, and, on the other, a "context-dependent assignment of informational content" that is required to safeguard informational pluralism (Allo 2007, 675). Allo examines different logical solutions to this problem but remains non-committal about them and, finally, ends up assuming that a message holding a situated informational content can have this content defined as a "situated constraining content" i.e., "it expresses a connection between (sets of) situations." Then "it functions as a constraint," a metalogic rule. It is "a message expressing a regularity. If true, it informs the receiver of how some situations carry information about others." (Allo 2007, 682 passim) The description of how a situation carries information about other situations is precisely the description of how a channel works. Following Barwise's suggestions, Allo's conclusion is that "a pluralist reading of a relevantly based account of informational content (as presented in this paper) mirrors and extends a suitably constructed reading of channel-theory." (Allo 2007, 686).

Allo's statement about the convergence between informational pluralism and theory of channel[6] does not regard only the relationships between logic, semantics, and information theory. It is not – or, at least, it should not be – an exclusively logical business. Barwise, Mares, Restall, have – along with Allo himself – shared a plan to defend a naturalistic view of information, designed to have a broader significance.[7]

Basically, the well-known Routley and Mayer ternary relation (R&M) permits us to develop the logical scheme of a *channel*, while interpreting the logical consequence as a *constraint*.[8] Barwise has argued that there is a correspondence between logical consequence and channel (Barwise 1993). They both relate situations to one another. Consequently, they also relate information contained in those situations and then put constraints on information flow. In more formal terms, the conditional $\phi \rightarrow \psi$ is a constraint. Hence, the entailment $c \vDash (\phi \rightarrow \psi)$ is true iff for each situation s_1, s_2, where $s_1 \longmapsto_c s_2$, if $s_1 \vDash \phi$, then $s_2 \vDash \psi$. So, "a channel $[\longmapsto_c]$ supports a constraint, just when for each pair of situations s_1, s_2 related by the channel $\longmapsto_c$ if s_1 supports the antecedent, then s_2 supports the consequent. This is the crux of information flow."[9] In a slightly different key, the ternary relation $s \sqsubseteq_c s'$ says that information available in situation s can be combined with the information available in c, to provide information available in s'. We may also define a particular set of situations $A_c\sqsubseteq$, about which we can gain information by combining the information that A with information available in c (Allo and Mares 2012, 174).

At this time, a philosophical consideration is in order. Assuming that the information we are talking about is composed of meaningful (or model-theoretic) data connected through the fragment of natural deduction system of relevant logic (Mares 2010, 117), we have to figure out what kind of information gains we can obtain. I am afraid the gains based on the above procedure are just semantic. They may also be model-theoretic, but it is not granted at all that they mirror physical phenomena until a verification process is achieved. This is why Mares takes a strong commitment to untie information from the exclusive logical and semantic status it has maintained. He insists on "objective" information and "real use" of it. Real use seems to be a successful extraction of the information available in an environment,

---

[6] Any theory of channel cannot but be connected to the information theory that Claude Shannon established in 1948.

[7] In the second part of this paper, I intend to show that informational pluralism may also impact the philosophy of science.

[8] See also Mares' relation of "accessibility" in Mares (2010), 117, and 125.

[9] Restall (1995), 465. See a few more details about the connection between the R&M ternary relation, the concepts of channel and code, in Sect. 6.4, below.

and it cannot be "explained" by a model theory in its typically semantic terms (Mares 2010, 123–124). Thus, Mares supports an *inversion* of the explanatory order: information has to explain meaning as well as consequence, not vice-versa (Allo and Mares 2012, 172). His purpose is seemingly to provide a naturalistic or physical foundation of the logic of information, which is made very difficult by the endless debates about the nature of the information.

In any case, information is not a logical entity. It is encoded in physical signals and is computed and transmitted through physical channels. In other words, information must be encoded, and the symbols that are part of every code are also physical items, transmitted via a physical channel. Suffice it to mention the dichotomy analog-digital, affecting both codes and channels. Consequently, information shares the physical character of computations (Piccinini 2015). Also, information consists of data that are part of an environment (Mares 2010; Allo and Mares 2012; Barwise and Seligman 1997). Allo, Mares and Barwise have accepted a correspondence between logical constraints and channels but have treated it exclusively in logical terms, failing to reckon with the puzzling intertwinement between their physical character and their logical use in computations. This is a problem both for information and informational pluralism.

So far, semantic information has been considered in logical terms. But we are now working with computational models where *the definition of meaning is achieved through different programs and programming languages* (O' Donnell 1998). Such programs are run by different kinds of semantics for programming languages. The classic paper by Van Emden and Kowalski, for one, enumerates three kinds of semantics that encode under different logical and computational systems the data to be packed into computational types. The structures of such semantics consist of abstract terms, formal relations, procedures, and functions, which are encoded in *formal metalanguages*. First, we have an *operational* semantics that defines the input-output relations computed by the program inside the machine. Second, we have a *model-theoretic* semantics that determines the denotation of symbols in a set of clauses and the entailment relations between those symbols. Finally, we have a *fixpoint* semantics that defines a minimal fixpoint for the transformations associated with the procedure of the definition of a term. The three semantics are mutually equivalent (cf. van Emden and Kowalski 1976; see also Fitting 2002). Such multi-layered architectures are built through intertwined informational and computational layers and, therefore, supported by inextricably dense connections between information and computation.[10]

If informational pluralism has to make sense of these multiple intertwinements between information and computation, it needs a more sophisticated (or powerful) semantics than one that can be exclusively set by logic. We must adopt a "situation semantics," which – following Barwise suggestions – treats meaning in terms of information conditions, not truth conditions (Mares 2010, 112). We need to go beyond the information that "encapsulates" truth. We have to work with

---

[10]Oron Shagrir has written that "Computation is information processing."

computational models, and then we have to craft a notion of information that may include an adequate mechanism, up to computing on the different languages in which are expressed the disparate data types processed in a model. This mechanism is *encoding* and has to be activated at various levels – logical, semantic, and computational. In this paper, I cannot provide a thoroughly detailed study of coding and its essential relationship with the overall phenomenon of the information. But I can show how it improves the logical status of informational pluralism.

I think that Allo, Mares, and Restall have correctly argued the concept of informational content and informational semantics, and have reached the limit a logical inquiry could reach, namely the interpretation of logical consequence as a transmission channel. They have also secured the concept of informational pluralism, under the premise that a relevant logic was in place. Unfortunately, Barwise ignored the concept of code and the necessary connections between it and the concepts of channel and information flow. Hence, Barwise analysis remains in the realm of logic and cannot complete the passage from logical pluralism to informational pluralism. Likewise, as long as Barwise overlooks the concept of code, he cannot obtain the benefits of information theory, which introduces a mechanism of its own – the compression of information – to explain how generalities are disseminated in distributed systems.

6.4   Coding

Codes are ubiquitous in all areas of human cognition and communication, and their unquestionable multiplicity is a straightforward vindication of pluralism. Both our language and our numerical system are codes, based on the twenty-six-letter alphabet and the decimal numeral system.

The theory of codes stems from Shannon's theory of information (Berstel and Perrin 2010, 7). Shannon focused on the manipulation of digital symbols, to the purpose of transmitting and storing of messages. A message consists of strings of encoded symbols. Shannon's well-known scheme for the transmission of messages shows that we process informational data by switching from source-codes to channel-codes (Shannon 1998, 34 ff.). We compute on data and re-code them to reach an optimal compression rate and maximize the channel capacity, i.e. "the efficiency of the coding system" (Shannon 1998, 62; Hankerson et al. 2003; Cover and Thomas 2006). Therefore, coding (in particular, digital coding) emerges as an essential component of information (Abramson 1963; Hamming 1986).

If Shannon's theory of communication represents the external shell of encoding, codes have an internal structure that primarily accounts for their power. A code is an algebraic semigroup, an alphabet, a monoid, made of symbols, and rules for the connection and transformation of symbols. Encoding "is the studying of embeddings of a free monoid into another." (Berstel and Perrin 2010, 7) To embed the elements of a monoid into another, we need a set of maps, $f: X \to Y$ that put in correspondence couples of elements $x_1 \ldots x_n$ and $y_1 \ldots y_n$ with each other. The mapping rules are arranged by a ternary relation that associates different correspondence rules to various couples of elements. Such correspondence is basically the R&M-type "channel" $\longmapsto_c$ that has been analyzed by Restall (see previous section). In other words, this is a morphism F between a type X and a type Y (Restall 1995). We say this is a code, a tool that operates real flows of information. A code is not a purely logical embedding but a mix of different context-dependent and context-independent morphisms that are managed by R&M ternary relations. Before being activated, a code is just an inert algebraic semigroup. Once it is activated, it becomes a tool that changes a message into another message of the same content but a different form. Thus, encoding is the primary support of informational pluralism.

6.5   Pluralism in Science

The more the debate unfolds, the more pluralism turns out to be a standard rather than an option. Not an easy one to meet, however. In general, to admit that the development of science is not a monolithic process is not to say that it is a pluralistic process. But informational pluralism, with the support of coding theory, can confront the complexity of the scientific method better than a logically-based semantic pluralism can do.

Woodward explicitly affirms his commitment to a "pluralistic understanding of science", to strike a balance between theory-driven and data-driven science (Woodward 2011, 171). But what kind of pluralism is this, and what does it amount to? Intuitively, pluralism can be identified by contrasting it with reductionism. But pluralism comes in different kinds and each kind is characterized by methods or doctrines that a pluralist approach puts together, precisely under the premise that the idea of a scientific method based on monistic logic is definitely behind us. So, pluralism can be broadly brought back to the need to revise the idea of scientific explanation and causation. Both mechanistic and unificationist explanations (see Woodward 2003, 350–375) presuppose some kind of pluralism. The use of informational concepts should help us to define a notion of "explanatory pluralism" that goes beyond the *de facto* pluralism so widespread in the philosophy of science. Let us consider, for example, the fragmentation of the concept of causation. In a retrospective paper significantly entitled "Causal Pluralism", Peter Godfrey-Smith remarks causation is an "essentially contested concept", and has become a controversial topic to be treated by every philosopher (Godfrey-Smith 2010). This has triggered further scholarship about causation. So, we can assume that causation "comes in at least two basic and fundamentally different varieties": counterfactual dependence, that can be computed stochastically, and singular production that turns out to be incomputable (Hall 2004; see also Strevens 2008 and Illari 2011).

The distinction between data and phenomena[11] has been the object of a series of papers by James Woodward and James Bogen. They explicitly say the upward inferences from data to phenomena do not follow only one but rather many avenues. This will prove extremely helpful in spelling out a pluralistic approach to the scientific method. The "Data-to-Phenomena Reasoning" (DPR) tells us "how we should understand the structure of scientific theories and the role of observation in science." Actually, Bogen and Woodward argue that there is "no principled distinction to be drawn between data and phenomena" (Bogen and Woodward 1988, 315) and that they "are inclined to be ontologically non-committal" about the status of phenomena (Bogen and Woodward 1988, 321). Hence, a definition of phenomena has to be drawn from a comparison with data and it will consist of epistemic features. Phenomena are objects of systematic explanations and predictions, provided by scientific theories that typically exhibit detailed patterns of dependency, under the form of either differential equations or causal mechanisms. Alternatively, patterns of dependency can be gathered under the form of more extensive systematic accounts that unify different scientific theories. What turns out to be more important than settling distinctions is to move upward from claims about data to claims about phenomena (Bogen and Woodward 1988, 324–326).

I will argue that Bogen and Woodward's "Data-to-Phenomena Reasoning" (DPR) provides an analytic pattern that can be combined with causal pluralism and makes room for informational pluralism. Obviously, fitting information theory into the pattern will require some facilitation. Ultimately, explanatory and informational pluralism amounts to the position that scientific claims are to be assessed in computational models.[12]

Before I provide a detailed analysis of the link between DPR and information theory, let me list a few features of DPR that show why this is a pluralistic method:

(1) first, experimental techniques that are employed for collecting data and (2) statistical techniques for analyzing them "seem to develop relatively independently of 'high' theories" (Woodward 2011, 170); (3) This is a premise to a more general statement concerning the autonomy of DPR from the theories that shape targeted phenomena (Woodward 2011, 171).[13] Finally, Woodward argues that (4) "top-down theory-dominated" views and "bottom-up" or "data-driven" reasonings are on the same par (Woodward 2011, 171). This is a critical point (see Sect. 6.3, above) because we are to reject the idea that we may have a unique scheme for designing causation patterns. "Efficient causation does not flow only upward, from system

---

[11]Data are recorded (and I think encoded!) observations. Phenomena are defined this way: "facts about phenomena are natural candidates for systematic scientific explanations in a way in which facts about data are not." (Bogen and Woodward 1988, 326)

[12]Bogen and Woodward do not consider models and simulations.

[13]More than this, I maintain that the pluralistic features of DPR also lead to the autonomy of computational models. I may even suggest that models, insofar as consisting of interpretive patterns applied to data collections, may be equated to the structures that connect and organize what Woodward calls "phenomena".

components to system behavior as a whole." Also, "downward causation" has to be taken into account (Bishop 2012, 9; Bogen and Woodward 2003).

"Data are effects produced by elaborate causal processes that may involve the operation of the human perceptual and cognitive systems as well as measuring and recording devices [...]. The production of data [...] requires the manipulation of highly transitory and unusual combinations of causal factors [...]. In contrast, phenomena are typically due to the uniform operation of a relatively small number of factors" (Bogen and Woodward 2003, 224). Therefore, there is a problem of data reliability assessment. Three points are to be considered: (1) Control of confounding factors; (2) data-reduction and statistical analysis; (3) Investigation regarding equipment (Bogen and Woodward 1988). What Bogen and Woodward are saying appears to be a confirmation of the thesis that all data are always encoded, and the role of observation in scientific research has to be revised: "The facts for which typical scientific theories are expected to account are not, by and large, facts about observables. Indeed, we think the whole category of observation [...] is much less central to understanding science than many have supposed." (Bogen and Woodward 1988, 305). To this effect, "Nagel appears to think that the sentence 'lead melts at 327 degrees C' reports what is observed. But what we observe are the various particular thermometer readings—the scatter of individual data-points." (Bogen and Woodward 1988, 308–309). Consequently, "while the true melting point is certainly inferred or estimated from observed data, on the basis of a theory of statistical inference and various other assumptions, the sentence 'lead melts at 327 ± 0.1 degrees C' [...] does not literally describe what is perceived or observed. [...] Data are idiosyncratic to particular experimental contexts, and typically cannot occur outside of those contexts." (Bogen and Woodward 1988, 317).

Woodward and Bogen mostly use the concept of information in its commonsense meaning, as a synonym of knowledge data. This is not all, however. They discuss information *theory* only once, in their 1988 paper, and the examination is much more careful. They dissent from Dudley Shapere's analysis of solar neutrinos. Given the weak interaction of neutrinos with other bodies, even pretty dense ones, they freely traverse vast regions of space and deliver "information" to an appropriate receptor. Shapere seemingly believed that such an information-theoretic setup might certify neutrinos as true observation carriers (cf. Shapere 1982). Do they convey credible "information" about the energy-producing processes occurring in the core of the sun? Do they provide "direct observation" data about solar core? Do such information-data count as an explanation or justification of phenomena occurring inside the solar core? Bogen and Woodward's answer to all questions is definitely negative. Moreover, Shapere's informational language offers them an opportunity to get to grips with information theory. They are aware that Shapere's attempt to smuggle solar neutrinos' experimental data as accurate observations or perceptions is deceiving, in various respects. First, it conflates data and phenomena. Second, mistakes information as a possibility of legitimating perceptual data. "Shapere has not so much a theory of observation or observation sentences as a promissory note we do not know how to cash." (Bogen and Woodward 1988, 348).

Most interestingly, Bogen and Woodward are also aware of the non-semantic character of Shannon's theory: "The only well-worked-out theory of information is the Shannon & Weaver theory which provides a mathematical characterization of information [. . .] in which the semantic content of the signal is irrelevant. There is nothing in the Shannon and Weaver theory to tell what counts as information in the sense of content conveyed by a signal. But this is the notion of information required for a theory of perception." (Bogen and Woodward 1988, 348) Consequently, any misleading identification of information with data must be avoided.

Bogen and Woodward's conclusion is that, on the one hand, observation treats just data and cannot justify phenomena or complex causal structures. On the other, information cannot justify observation. Should it be possible that information justifies phenomena? Indeed, it should not. This paper is not about epistemology and justification but rather about semantics. Accordingly, I suggest another perspective: information may travel upward and take part in the process of encoding data and modeling phenomena. Let me say it again: modeling is different from justifying. I am not after realism but encoding. In general, information can supervise the acquisition and handling of statistical data that are available in databases. In the following, I will explore such a possibility. Information theory can take part in various operations, such as computing types, checking errors, testing hypotheses, estimating parameters (Cover and Thomas 2006, 347, and ff.) It may also have a part in the process of manipulating causation (Woodward 2003). However, I am not going to pursue such subjects.[14] I will instead focus on the concept of relevance and examine the relationship between relevance logic, pluralism, and the statistical-relevance model introduced by Salmon.

6.6 Information and Pluralism

Let me remark that once relevance logicians admit incompleteness and partiality of their perspective, along with the impossibility of deploying a calculus capable of formulating cogent decisions, the contribution of a statistical-relevance model for handling data cannot but be welcome. Shannon's information is indeed a tool for evaluating probability rates and statistical relevance (Shannon 1998, 59). The primary object of information theory is the design of encoding systems. As I mentioned above, I disagree with Allo, on this point. He holds that semantic information (true meaningful data) is all of information and – apart from the question of subjective probability connected to personal epistemology of "being informed" – pays little attention to information as an elaboration of objective probability. Among relevance logicians, Edwin Mares provides a more detailed treatment of this issue. It will be helpful to consider his approach below while reviewing in some detail Salmon's stance.

---

[14]For an epistemic synthesis of Woodward's ideas about causality, see Gonzalez 2018.

On the one hand, there is certainly a substantial difference between relevant logic and statistical relevance regarding their procedures. On the other hand, there is an apparent difference between logic and statistics regarding their purpose. Nevertheless, there is also a basic common purpose: if relevant logic aims to provide a sufficient tie between an antecedent and a consequent in material implication (Mares and Meyer 2001, 283), Salmon's statistical-relevance (SR) model pursues this aim as for the relation between cause and effect (Salmon 1984).

As for the problem of our interest, the shared purpose consists of circumscribing and identifying partially ordered sets of either homogeneous *reference classes* (Salmon) or *compatible cases* (Allo) or mutually *accessible situations* (Mares). They may work as backdrops to information-theoretic analysis, as a replacement for both classical logic and the deductive-nomological approach. So, relevance logic can be used to provide a logical and ontological description of relations that lie behind the statistical homogeneity of reference classes.

The expressions that are used by Salmon and Allo belong roughly to the same kind. Salmon traces back his "objectively homogeneous reference class" to Richard von Mises' term "collective," which "denotes a sequence of uniform events or processes which differ by certain observable attributes, say colors, numbers or anything else." (von Mises 1981, 12) I think that also Allo's concept of "Star Information Frame", which tries to put together *compatible* events, can be included in this *genus*. Both *homogeneous* and *compatible* events are not rigorously uniform. They are only partially uniform. Moreover, homogeneity may count as a generalization of the "variable sharing constraint" (Mares and Meyer 2001, 283). Homogeneity is determined by several chain relations retrieved between propositions and situations. A "variable sharing constraint" (Mares and Meyer 2001, 283) is a necessary but not sufficient criterion that certainly helps in determining the requested homogeneity. More accurately, this job can be done through the well-known ternary relation R between situations and pairs of propositions. Hence, for situations a, b, c, $Rabc$ occurs iff for all propositions X, Y, if information $IaXY$ is true and b is an element of X, then c is an element of Y. Under these circumstances, the material implication $A \to B$ is true at the situation a iff for all the situations b and c, the truth at b of $Rabc$ and A, entails the truth of B at c (Mares 2006, 402). In this line of reasoning, Mares provides further logical structures, such as "R-frame" and Lattice of Subsets, conceived to the purpose of increasing the mutual relevance of different situations included in a class and the reciprocal links between their probability functions (Mares 2006, 403–405; see also Mares 1997). In addition, Salmon has appropriately argued that the concept of reference class is to be refined through Church's concept of effective computation on random sequences.

The fundamental payoff of the work that has been done so far on relevant logic and information is this: the logical relations of accessibility and compatibility established between situations permit them to be encoded either under the form of *phenomena* (*sensu* Bogen and Woodward 1988), or in a given "level of abstraction" (*sensu* Floridi), or in the framework of a computational model. They can also be recoded for being transmitted from one version of a model to another. Transmissibility can be assumed as a pragmatic version of *partial* generality or an operational basis for the construction of regularities. Furthermore, this is precisely the contribution of relevance logic and information theory to the philosophical problem of pluralism.

6.7 Statistical Relevance and Information

As I mentioned above, Salmon's Statistical-Relevance model (S-R model) provides an analysis of statistical data of the kind considered in Woodward's "Data-Phenomena Reasoning".

Salmon's equivalence between transmission of information and propagation of causal influence has proved not to work as a substantial theory of *physical* causation. But this does not prevent information theory from being valid support for an "ontic" conception of scientific explanation and *manipulation* of causal agencies. I intend to show that the use of information theory benefits from Salmon's philosophical support, and it turns out to be convergent with James Woodward's analysis of manipulation and intervention. It is also to be stressed that the connection between information theory and causation which is being established via Shannon's information, Barwise's "information flow" (Barwise and Seligman 1997) and "semantic information theory" (cf. Floridi 2011) will vindicate the overall capability of information theory to deal with questions in the scientific method. This idea has been downplayed for a long time and I hope it will be revised, sooner or later.

The pluralist character of the overall process is secured via the concept of logical relevance analyzed throughout this paper. We have shown that Routley and Meyer's semantics (R&M semantics) is an *intrinsically pluralistic* structure insofar as it connects a finite number of different situations in a framework. Then, they can be stored in an appropriate database.

An S-R model consists in detecting causal relevance between a number of given events or "situations". Probabilistic dependencies are measured, and such measures are considered as an informational content of events or situations.[15]

We have seen relevance logic deals with *situations*. Situations can be parts of Routley and Meyer's "worlds." R&M semantics gives a way of understanding situated inference. Situations can be considered as containers of information about *actual* worlds. *Internally*, situations hold only partial information about worlds but provide relevant relations between data. They constrain data and relations between data. When such relations are probabilistic, we need to develop an analysis of probabilistically situated inference. To this effect, we need to add a relevance parameter r to a material implication so that $\rightarrow$ is replaced by $\rightarrow r$. Hence, "our problem is now to integrate a semantics for probabilistic relevant implication into the Routley & Meyer Model theory for relevant implication." (Mares 2006,

---

[15]This is based on the well-known equivalence – asserted by Claude Shannon (1998, 40) – between a discrete information source and a stochastic process.

402) Such manipulations yield information about mutual implications that can be detected inside the data set. This comes close to Salmon's notion of "homogeneity of reference classes" or data homogeneity.

Externally, situations must be mutually accessible ("compatible" in Allo's words). From this point of view, they can be deemed as "abstract entities [which], like properties and other *abstracta*, can exist in more than one place at one time." (Mares 2006, 401) We can use our power of abstraction, as is now usual in computational science, to highlight certain features of a body of knowledge, for example, "causal connectivity" (cf. Levy and Bechtel 2013) particularly important in the present analysis. In other words, to recode the data contained in certain situations, we use, at a given time, the logical and informational relations between the symbols we have used to encode other data, in other situations, at a previous time. Data are repeatedly fitted into the patterns of a model. Such multiple operations of revision and recoding are typical of computational models (see Padmanabhan et al. 2013; Leonelli 2008). We perform logical abstractions, treat classes as situations,[16] and use the above-described tools of relevant logic to manipulate those abstractions. As a result, we may have a new and more detailed discernment of the internal structure and external relations of situations.

In an S-R model (Salmon 1984, 32 and ff.), a set of factors $C_{1\ldots n}$ affects the occurrence of a set of probable events $B_{1\ldots n}$ under the circumstances $A_{1\ldots n}$. Such circumstances are the reference class for us to take apriori probabilities $p(B_i|A_i)$ that B will happen. As a follower of Bayes,[17] Salmon proceeds assuming that the apriori probabilities will be repeatedly modified by the posterior impact of some combination of factors C over events B. The purpose of the S-R model is precisely to measure such impact and the resulting posterior probabilities. Therefore, we partition the reference class and select an *explanandum partition* out of the initial reference class according to ordering criteria derived from our preliminary causal knowledge. This results in a set of attributes $B_{1\ldots m}$ over B. Then, we outline a kindred *explanans partition*, by selecting a set of statistically relevant factors $C_{1\ldots s}$. The set of all those relations is what Salmon calls an *S-R basis*. We put together the probabilities of the two partitions[18] and have a number of probabilities combinations $p(B_{1\ldots m}|A. C_{1\ldots s})$ that constitute a number of mutually exclusive and exhaustive cells. They can be re-ordered in a states-space or a matrix such as those used in data analysis. A statistical explanation consists of modifying a probability estimate through a recalculation and thereby a relocation from one cell to another or from one matrix to another. If such a process of recalculation implies recoding operations, it consists of a transmission of information and therefore we have an "information-theoretic" version of epistemic explanation. In analyzing the information-theoretic

---

[16]Besides the equivalence between situations and classes, we may recall other important computational equivalences, such as that between propositions and types, which is the basis of the fundamental Curry-Howard Isomorphism.

[17]See Salmon (1967), 58–62 and 108–131; see also Salmon (1970).

[18]I am slightly departing from Salmon's account at this point.

explanation, Salmon uses the toolkit of information theory. He also calls for the use of computability theory.[19]

As a result, an explanation can be deemed as analogous to the transmission of information. Salmon argues that an S-R model measures "the information transmitted in any explanatory scheme" (Salmon 1984, 38) and, conversely, "a good measure of the value of an S-R basis is the gain in information furnished by the complete[20] partitions and the associated probabilities. This measure cannot be applied to the individual cells one at a time." (Salmon 1984, 41). "Instead of regarding the goal of a scientific explanation as the achievement of 'nomic expectability' on the basis of deductive or inductive arguments, one can look at scientific explanations as ways of increasing our information about phenomena [. . .]. We can borrow the technical concept of information transmitted from information theory, and then *go on to evaluate explanations in terms of the amounts of information that they transmit* [my emphasis]. Since 'information' is defined probabilistically in information theory, this idea is especially well suited for dealing with statistical explanations." (Salmon 1984, 97) Such a gain in information allows us "to evaluate explanations in terms of the amount of information that they transmit." (Salmon 1984, 97) How do we measure this? We have to recall that a gain of information is equivalent to a reduction of uncertainty and we can, therefore, measure − using "standard information-theoretic concepts"[21] − the uncertainty H of both explanandum and explanans partitions {M} and {S}:

$$H\,(M) = \sum -p_i \log p_i$$

$$H\,(S) = \sum -p_j \log p_j$$

---

[19] Salmon was aware that computability theory was needed for the crucial problem of the construction of objectively homogeneous reference classes (see Salmon 1984, 58–60, 67–68, 81). As for information theory, see Salmon 1984, in particular, 97–101, 125–126, 139–154. S-R approach has been recently updated in informational terms, especially via the notion of "causal power". This represents the degree to which changes in a cause C produce changes in an effect E. Causal power uses relevance logic to cover not only conditional probabilities and statistical datasets but also Bayesian Networks used in Artificial Intelligence. Causal power is measured using Shannon's entropy concept (Korb et al. 2009).

[20] The completeness of partitions provides to Salmon's S-R models the character of objectivity that distinguishes them from Hempel's I-S models (Salmon 1984, 41). Completeness is granted by the maximal homogeneity of both explanans and explanandum partitions (Salmon 1984, 37). It is also to be mentioned that Salmon's objective homogeneity is the same thing as Shannon's "ergodicity," roughly defined as "statistical homogeneity" (Shannon 1998, 45), and reputed to be essential for the final Theorem 11 of Information theory (cf. Kinchin 1957).

[21] Salmon's and Greeno's visible source is Kullback's classical treatise *Information Theory and Statistics*. Kullback supplements Shannon's concepts with the works of Fisher on information (cf. Kullback 1968). However, apart from Salmon's theory, it is crucial to shift from Fisher's to Shannon's view. In doing so, we gain an intensional perspective on statistical types that is the premise for being able to build computational models. Sayre also misses this point.

The uncertainty of the overall system is

$$H(M \times S) = \sum_{i=1}^{m} \sum_{j=1}^{s} -p_i p_{ij} \quad \log \ p_i p_{ij}$$

and the transmitted information I_T that can be considered equivalent to the gain of information provided by a scientific explanation

$$I_T = H\ (M) + H\ (S) - H\ (M \times S)$$

There is little doubt that Salmon's statistical-relevance approach can support the centrality of computational models in present-day science. To this effect, I have shown the affinity of the S-R approach and informational pluralism in the field of data modeling. I also believe that Salmon's information-theoretic approach to scientific explanation may count as a philosophical warrant to take information as a code for analyzing causation.

A more challenging job should be to assess whether Salmon's information-theoretic version of scientific explanation might support synergy between informational pluralism and data modeling while remaining consistent with Salmon's realist strategy in the philosophy of science.

He argues that information-theoretic concepts engage and measure "the mechanisms that operate throughout the physical world" (Salmon 1984, 126; Sayre 1977). Thus, the survival of information-theoretic version requires further verification: "The information-theoretic version of the epistemic conception [of scientific explanation] survives just in case it devotes enough attention to the causal processes of information transmission (i.e., the transmission of causal influence) to qualify for a transfer from the epistemic to the ontic conception." (Salmon 1984, 133). Salmon has not explicitly granted any qualification. Quite to the contrary, under the pressure of critics such as Dowe, Kitcher, Cartwright, he has stressed the physical character of causation and sidestepped the analogy between causal propagation and information transmission (Salmon 1994). Actually, there is little or no room to reconcile pluralism with scientific realism.

Nevertheless, assessing whether information theory or informational pluralism is compatible with scientific realism was beyond the scope of this paper, from its inception. On the contrary, my goal was to investigate the nature of pluralism and the role of encoding – a *de facto* fundamental role – in scientific representation and modeling. With this aim in mind, Salmon's S-R and information-theoretical approach still hold their appeal.

6.8   Concluding Remarks

I have argued elsewhere (Camardi 2012) that information theory, along with the theory of computation, is an essential tool for the construction of computational models. If one inquires into the structure of scientific and computational models, a pluralist character, as has been specified above, is straightforwardly detectable in the cumulative flux of computational data and the resulting multiple revisions of explanatory mechanisms. Hence, the question referred to as "explanatory pluralism" – namely the intervention of many scientific theories in an explanation – is a part of this line of development. Must one refer to logical and informational pluralism in Beall and Restall, Allo, Mares, Barwise's sense? Alternatively, has one to refer to the current attitude of using a cognitive key in many kinds of explanation? To be sure, they are both exciting avenues. I have not pursued, in this paper, the cognitive avenue. I think, however, that explicit awareness of the primary role of encoding may serve as a bridge to link up the logico-computational and the cognitive level of research. Salmon's statistical-relevance model of explanation, with the visionary character of its informational component, has played a role in shaping this blueprint.

References

Abramson, N. (1963). *Information theory and coding*. New York: McGraw Hill.

Allo, P. (2007). Logical pluralism and semantic information. *Journal of Philosophical Logic, 36*, 659–694.

Allo, P., & Mares, E. (2012). Informational semantics as a third alternative? *Erkenntnis, 77*, 167–185.

Barwise, J. (1993). Constraints, channels, and the flow of information. In P. Aczel, D. Israel, S. Peters, & Y. Katagiri (Eds.), *Situation theory and its applications* (pp. 3–27). Stanford: CSLI.

Barwise, J., & Seligman, J. (1997). *Information flow: The logic of distributed systems*. Cambridge: Cambridge University Press.

Beall, J. C., & Restall, G. (2000). Logical pluralism. *Australasian Journal of Philosophy, 78*, 475–493.

Beall, J. C., & Restall, G. (2006). *Logical pluralism*. Oxford: Oxford University Press.

Berstel, J., & Perrin, D. (2010). *Codes and automata*. Cambridge: Cambridge University Press.

Bishop, R. (2012). Fluid convection, constraint and causation. *Interface Focus, 2*, 4–12.

Bogen, J., & Woodward, J. (1988). Saving the phenomena. *The Philosophical Review, 97*, 303–352.

Bogen, J., & Woodward, J. (2003). Evading the IRS. *Poznan Studies in the Philosophy of Science and Humanities, 20*, 223–256.

Bueno, O., & Shalkowski, S. (2009). Modalism and logical pluralism. *Mind, 118*, 295–321.

Camardi, G. (2012). Computational models and information theory. *Journal of Experimental and Theoretical Artificial Intelligence, 24*, 401–417.

Caret, C. (2019). Why logical pluralism? Synthese. https://doi.org/10.1007/s11229-019-02132-w. Accessed on 06 Sept 2019.

Cook, R. (2019). Pluralism about pluralisms. In J. Wyatt, N. J. L. L. Pedersen, & N. Kellen (Eds.), *Pluralism in truth and logic* (pp. 365–386). Cham: Palgrave Macmillan.

Cover, T., & Thomas, J. (2006). *Elements of information theory*. Hoboken: Wiley.

Dretske, F. (1981). *Knowledge and the flux of information*. Oxford: Blackwell.

Fitting, M. (2002). Fixpoint semantics for logic programming. *Theoretical Computer Science, 278,* 25–51.

Floridi, L. (2011). *The philosophy of information.* Oxford: Oxford University Press.

Godfrey-Smith, P. (2010). Causal pluralism. In H. Beebee, C. Hitchcock, & P. Menzies (Eds.), *Oxford handbook of causation* (pp. 326–337). Oxford: Oxford University Press.

Gonzalez, W. J. (2018). Configuration of causality and philosophy of psychology: An analysis of causality as intervention and its repercussion for psychology. In W. J. Gonzalez (Ed.), *Philosophy of psychology: Causality and psychological subject. New reflections on James Woodward's contribution* (pp. 21–70). Boston/Berlin: de Gruyter.

Hall, E. (2004). Two concepts of causation. In L. Paul, E. Hall, & J. Collins (Eds.), *Causation and counterfactuals* (pp. 225–276). Cambridge, MA: MIT Press.

Hamming, R. W. (1986). *Coding and information theory.* Englewood Cliffs: Prentice Hall.

Hankerson, D., Harris, G., & Johnson, P. (2003). *Introduction to information theory and data compression.* Boca Raton: Chapman and Hall/CRC.

Illari, P. (2011). Why theories of causality need production: An information transmission account. *Philosophy and Technology, 24,* 95–114.

Kinchin. (1957). *Mathematical foundations of information theory.* New York: Dover Publications.

Korb, K., Hope, L., & Nyberg, E. (2009). Information-theoretic causal power. In F. Emmert-Streiss & M. Dehmer (Eds.), *Information theory and statistical learning.* Berlin/New York: Springer.

Kullback, S. (1968). *Information theory and statistics.* Mineola: Dover Publications.

Leonelli, S. (2008). Performing abstraction. Two ways of modeling Arabidopsis Thaliana. *Biology and Philosophy, 23,* 509–528.

Levy, A., & Bechtel, W. (2013). Abstraction and the organization of mechanisms. *Philosophy of Science, 80,* 241–261.

Mares, E. (1997). Relevant logics and the theory of information. *Synthese, 109,* 345–360.

Mares, E. (2006). Relevant logics, probabilistic information and conditionals. *Logique et Analyse, 196,* 399–411.

Mares, E. (2010). The nature of information: A relevant approach. *Synthese, 175,* 111–132.

Mares, E., & Meyer, R. K. (2001). Relevant logics. In L. Goble (Ed.), *The Blackwell guide to philosophical logic* (pp. 280–308). Oxford: Blackwell.

Morrison, M. (2015). *Reconstructing reality: Models, mathematics, and simulations.* Oxford: Oxford University Press.

O' Donnell, M. (1998). Introduction: Logic and logic programming languages. In D. Gabbay, C. Hogger, & J. Robinson (Eds.), *Handbook of logic in artificial intelligence and logic programming* (pp. 1–67). Oxford: Oxford University Press.

Padmanabhan, K., et al. (2013). In situ exploratory data analysis for scientific discovery. In T. Critchlow & K. K. van Dam (Eds.), *Data intensive science* (pp. 301–350). Boca Raton: Taylor and Francis.

Piccinini, G. (2015). *Physical computation: A mechanistic account.* Oxford: Oxford University Press.

Restall, G. (1995). Information flow and relevant logic. In J. Seligman & D. Westerståhl (Eds.), *Logic, language and computation: The 1994 Moraga proceedings* (pp. 463–477). Stanford: CSLI Press.

Russell, G. (2019). Varieties of logical consequence by their resistance to logical nihilism. In J. Wyatt, N. J. L. L. Pedersen, & N. Kellen (Eds.), *Pluralism in truth and logic* (pp. 331–361). Cham: Palgrave Macmillan.

Salmon, W. (1967). *Foundations of scientific inference.* Pittsburgh: University of Pittsburgh Press.

Salmon, W. (1970). Bayes's theorem and the history of science. In R. Stuewer (Ed.), *Historical and philosophical perspectives of science* (pp. 68–86). Minneapolis: University of Minnesota Press.

Salmon, W. (1977). An "at-at" theory of causal influence. *Philosophy of Science, 44,* 215–224.

Salmon, W. (1984). *Scientific explanation and the causal structure of the world.* Princeton: Princeton University Press.

Salmon, W. (1994). Causality without counterfactual. *Philosophy of Science, 61,* 297–312.

Sayre, K. (1977). Statistical models of causal relations. *Philosophy of Science, 44*, 203–214.

Shannon, C. (1998). The mathematical theory of communication [1948]. Repr. In C. Shannon & W. Weaver (eds) *The mathematical theory of communication* (pp. 29–125). Urbana: University of Illinois Press.

Shapere, D. (1982). The concept of observation in science and in philosophy. *Philosophy of Science, 49*, 485–525.

Strevens, M. (2008). *Depth. An account of scientific explanation*. Cambridge, MA: Harvard University Press.

Turner, R. (2009). *Computable models*. London: Springer.

van Fraassen, B. (2008). *Scientific representation*. Oxford: Oxford University Press.

Van Emden, M., & Kowalski, R. (1976). The semantics of predicate logic programming language. *Journal of the ACM, 23*, 733–742.

Von Mises, R. (1981). *Probability, statistics and truth*. Mineola, New York: Dover Publications.

Woodward, J. (2003). Making things happen. In *A theory of causal explanation*. Oxford: Oxford University Press.

Woodward, J. (2011). Data and phenomena. A restatement and defense. *Synthese, 182*, 165–179.

Chapter 7
The Methodology of Theories in Context: The Case of Economic Clustering

Catherine Greene and Max Steuer

Abstract The search for laws in social science often goes hand in hand with appeals to universalism. This paper builds on the philosophical literature which holds that laws play little if any role in social science including in economics. We take this position as given for present purposes and go on to explore some implications of it for methodological universalism in economic enquiry. We note that a single methodology of economics is unlikely to be revealingly applicable to the variety of endeavours economists undertake, including foundational work, empirical investigation, policy enquiry, and training activities. We focus on empirical enquiry and argue that economic knowledge in this area progresses through back and forth interaction between observations and theories. The role of models in this process is explored. We argue that theories are revealing in context, though incapable of universal application. We draw attention to the *ad hoc* nature of much economic enquiry and illustrate and defend a pluralist approach to economic methodology using the example of economic clustering. Finally, we distinguish this theories-in-context approach from simple common sense.

Keywords Laws · Economics · Methodology · Clustering

7.1 Introduction

Our primary purpose is to describe a pluralist methodology of one aspect of economic enquiry and to argue for the appropriateness of our description. We call this approach 'theories in context'. This methodology is applicable to a wide class of economic endeavour but not to many other activities in economics. It is not always appreciated that economists apply economic analysis in a number of

C. Greene · M. Steuer (✉)
London School of Economics, London, UK
e-mail: m.steuer@lse.ac.uk

© Springer Nature Switzerland AG 2020
W. J. Gonzalez (ed.), *Methodological Prospects for Scientific Research*, Synthese
Library 430, https://doi.org/10.1007/978-3-030-52500-2_7

quite distinct, though related, activities.[1] Among these activities are foundational enquiries, often concerned with the logic of choice and preferences, both for individuals and for collectives. Experimental work on choice and related matters extends the purely analytical investigation of choice and preferences in the direction of psychological considerations. These investigations are only distantly related, if at all, to economists' many policy concerns such as: whether or not to build a third runway in the United Kingdom and, if so, where; is Greece better or worse off in the Euro; and how to regulate banks. Yet another concern is educational. Much of economic analysis is directed specifically at the training of economists and is focussed more on acquiring appropriate methods of thought than on propounding theories of the behaviour of economies, or methods for testing these theories, except as training exercises. It can be debated whether this training is appropriate for most of the work economists will undertake. Few observers would claim it should play no role, while a number question its dominance in the mix of possible training approaches. The point here is that any universal methodology intended to apply to all of these activities is likely to be superficial at best and misleading at worst. This is one form of methodological pluralism- pluralism that arises from the different economic activities of economists. We want to go further though and argue for pluralism at the level of empirical economic enquiry.

Two stumbling blocks in discussions of the methodology of economics are the existence, status, and significance of laws of economics in explanations of economic phenomena, and the use of patently unrealistic assumptions as part of an economic analysis. The possible justifications, or criticisms, of employing laws and unrealistic assumptions depend entirely on the particular kind of economic activity undertaken. In this discussion we concentrate exclusively on economics as an empirical social science which is concerned with understanding particular economic phenomena. 'Theories in context' is intended to apply to empirical studies with causal implications. It relates distantly, if at all, to engineering approaches to stabilising economies, to simulation and to other non-causal empirical investigations.

A substantial literature addresses the question of generalisation in social science.[2] On balance the consensus is that there is little in the way of successful generalisation, or laws, in social science. Some writers are optimistic regarding the future, and argue that the lack of substantial success in finding laws so far does not mean that researchers will not formulate successful generalisations in the future. Other writers see social concerns and the subject matter of social science enquiry as having certain characteristics which by their very nature result in a very limited role for generalisations or what are commonly called scientific laws. Without going into the arguments, for present purposes we adopt the position that empirical investigation in economics does not proceed through economic laws.

---

[1]For example, Hausman (1992) where economics is generally treated as being a single activity. This can lead to a rather abstract notion of the 'essential' nature of economics.

[2]See, for example: Brown (1984); Dray (1957); Little (1993).

Though starting from a position that much of economic analysis does not rest on laws, we have to concede that the two most widely quoted practitioners in economics writing on methodology, Lionel Robbins (1935) and Milton Friedman ([1953] 1966), appear to see a central role for economic laws, though in very different ways. We use the cautious qualification 'appear' because the references to laws may be suggestive rather than literal. Both expositions have retained interest because they are rich and complex undertakings. A brief observation cannot do justice to them. However, it is fair to say that Friedman sees economics as much like natural science, perhaps with an instrumentalist twist. He claims that economics has laws which have implications which can be tested. Robbins sees economics as having laws, but holds that these laws cannot be tested. The basic reason for this is that economic laws, while real and operative, are embedded in complex social interrelations which make them near impossible to observe. These Robbinsian laws arise from the application of logic to a few facts. These facts are found largely through introspection. The appeal of laws in economics is that if we find them, they will tell us what will, probably, happen in similar situations. As such, the hope of finding laws goes hand in hand with universalism, and the expectation that well-defined methodological strategies will be defined over time, which can be applied in future situations. We argue instead that enquiry in this important class of economic activity can best proceed with methodological variety, a plurality of strategies, and a certain amount of pragmatism.

7.2 Theories in Context

One obvious alternative to general laws is theories which provide reasonably accurate accounts of dominant or important causal relations in particular settings. The first question to ask is how do we know, or what are the grounds for believing, that the candidate explanation is a reasonably good theory of what is going on in the specified situation? The second question is if this is a relevant theory, why might it not apply in all similar situations?

Confidence in a theory of a particular event builds up gradually based on an iterative process of observation and explanation interacting with each other. Intuition may lead from observation to initial theorising. More developed theory in the form of a well specified model can lead to more care in observation. This process also has an historical aspect as a range of related observations and theories build up over time. Of course, economists cannot be sure that the explanation held to be applicable in a particular context is in fact broadly reasonable. A major task for economists is to predict the consequences of policy initiatives. A fairly accurate understanding of the causal factors at work in a particular situation is central to this activity. Opting for a particular theory is bound to be a judgement call, but one which is based on careful analysis.

A good working theory of a particular economic event is very likely to have some validity in explaining a similar event. If there is no discernible difference between two events, the candidate theory is equally applicable in the new event, yet to be studied in its own right. This matter of logic does not get us very far in actual applications because of uncertainty regarding similarity. All the significant factors may be the same this year as last year, but this year's event is a repetition, and that in itself can change responses. As a practical matter, given the amount of continuous change in the economy and the society in which it is embedded, at the very least some care and additional investigation is usually needed when applying a theory in a new setting.

A general statement of a theory is different from an explicitly specified model. Philosophers of economics have sometimes exhibited unease about the role of models in the process of economic discovery. Much of the current literature displays considerable uncertainty about the relation between models and theories (Morgan 2010). This apparent confusion is hard to justify as the relation between theories and models is a fairly simple matter. A model is a specific version of a more loosely and generally stated theory. For example, the theory may contend that the search for the best price results in some outlet clusters. But that is far from a detailed explanation of how these outlet clusters come about. A model spells out in explicit and unambiguous terms the explanation offered by the theory. Many different models can serve as specifications of the same theory. They may be trivially different or have significant differences. As a social scientist one may hold that a particular theory of clustering is correct in a certain context, without adhering to a particular model, or specification, of the explanation. Alternatively, one may lean towards a particular specification. We illustrate the concept of theories in context, and the role of models in that process by taking the case of economic clustering.

7.3 Economic Clustering

Economic clustering is partly a spatial phenomenon, but clustering also refers to the structure of product offerings and the similarity of products in the market place. Another example of clustering occurs at the production level, where independent producers choose to locate near to each other. Normally, where there is clustering there also are outliers who operate outside of the cluster. Economists are interested in clustering because it suggests that forces are at work which bring about the clustered structure and, while these forces are interesting in their own right, they can be important for economic policy. The presumption is that in the absence of forces producing economic clustering we would observe a more random distribution of economic activity both in the spatial sense and in the sense of product space.

It is important to recognise that frequently in economics there are many theories that can potentially be applied to an economic phenomenon. We do not have one theory of unemployment, but many theories. And so it is with economic clustering. In order to explain our view of one aspect of economic method, we begin by setting out a range of theories of economic clustering. Perhaps the best known of these is Hotelling clustering named after its major exponent (cf. Hotelling 1929, 41–57).

This kind of clustering can occur in both a geographic space and in product space. The central elements in Hotelling clustering are producer desire to maximise market share and consumer propensity to gravitate to the supplier who most closely matches their preferences. The match need not be exact. Consumers are assumed to choose among the goods or outlets on offer, and select the one closest to their preferences. A familiar abstract example is that of two ice-cream sellers on a beach. The product and price are assumed the same for both sellers. Consumers are spread evenly, more or less, along the beach. They prefer to buy from the nearest seller. This results in the two sellers moving close to each other. Moving closer to the rival seller increases market share for the mover and reduces that of the seller being closed in on. The move results in picking up some of the buyers who used to be closer to the rival, without losing any customers who now have to travel further but still find the same seller to be the closer.

With more than two sellers, no seller wants to be caught in the middle between two sellers. If trapped in the middle of the cluster, she will move out, but only far enough out so that there is no seller on one side. The cluster of sellers is not static. There is constant reshuffling, sometimes called leapfrogging, but the cluster never spreads far out. It will tend to concentrate in the centre of the market. This example with geographic space can readily be expanded to cover the types of product on offer in the market, rather than the location of identical products. People will have preferences for different sizes of car, different degrees of sweetness of soft drinks, different colourfulness of clothing, different degrees of strength of cleaning material, and so on. By going to the edge of the cluster one captures the market of all consumers with preferences just beyond the nearest supplier. To go any further in product space means losing some customers to the nearby supplier without picking up any more from elsewhere. So in many areas we might expect product clustering as long as there is not an enormous number of suppliers.

Economists have studied Hotelling clustering in a great variety of settings. One example is the clustering or airline departure times. Leaving just a bit later or earlier than the existing carriers helps to attract either all who would prefer an earlier flight or all who would prefer a later flight. Moving farther than a bit away from the others in departure time risks losing some passengers to the rival companies. Some passengers might prefer an earlier flight, but not that much earlier. Going to the edge of the cluster is usually better for market share, and accounts for the existence of the cluster (cf. Borenstein and Netz 1999).

Television and radio broadcasting provide compelling examples of Hotelling clustering. Broadcasters usually want to maximize their audiences. Viewers and listeners are often drawn to the activity of viewing or listening, and while they may have particular preferences, will choose the offering closest to their preference even if the match is not exact, or even if it is not very close. This provides fruitful ground for Hotelling analysis of clustering (cf. Steiner 1952, 1961). We will return to this example later.

Hotelling style analysis has obvious applications in the political sphere. Political parties have every incentive to maximise their appeal to voters. No doubt ideological concerns shape policy, but at the same time parties are inclined to position themselves in a way that attracts voters and potentially makes them electable. In a two-party system this can lead to convergence. Being slightly to the left, or the right, of the rival party, depending on orientation, is vote winning. Being further along the spectrum, one way or the other is to risk losing votes. The assumption being that voters will go for the party that most closely fits their views. When political parties have strong ideological positions which they are not prepared to abandon, or modify significantly, the wish to attract voters by strategic location of the policies is largely set aside. A less ideological party has an incentive to move closer to the other contender, picking up some votes from it, and not losing any other voters, again on the assumption that voters will go for the party that more closely matches their political views (cf. Downs 1957; Osborne 1993).

While Hotelling clustering is seen in many situations, it is far from providing the only model of the configuration of forces that leads to a cluster. Clusters can come about for a great variety of other reasons. One context in which we see clustering is in some types of retail outlets. A familiar example was the large number of electronic goods stores on Tottenham Court Road in London. This cluster has largely disappeared. That may be due to internet buying. Where Hotelling clustering depends on travel costs, both in geographic and in product space, some other clusters are driven by information costs. When consumers are unsure where to find the products—unlike ice-cream buyers on the beach—and also are unsure about which product to buy, a cluster may arise. Sellers may have higher rents in the cluster, and may face stronger price competition there. But by locating in the cluster they have a greater flow of customers for two reasons. It may be harder for consumers to find suppliers outside the cluster, and it may be more difficult to compare products and prices away from the cluster. This cause of the cluster is essentially consumer information (cf. Wolinsky 1983; Asarni and Isard 1989). Acquiring information is costly, and the cluster can reduce this cost to consumers. This is quite different from Hotelling clustering where information, or the cost of it, plays no role.

A related, but somewhat different, source of clustering is due to quality signalling. Consumers may know they want a doctor, for example, but feel they do not know how to judge the ability of a doctor. An urban location with a reputation of superior quality is a different kind of cluster. The clustering of theatres and art galleries may also signal quality. Consumers may read reviews, but they do not have direct knowledge of the quality of a theatrical offering. If a theatre locates in what is known as a quality zone, that potentially signals quality. This consideration can lead to yet another type of cluster (cf. Christou and Vettas 2005).

The theory of clusters has something to say about which products are on offer. Much of the analysis of markets starts and ends with consideration of a particular product, and with no discussion of why that product and not some other product has been put on the market. While the normal economic analysis of supply and demand does not specify which products are on the market it does set bounds. The theory eliminates from the market all products where the supply price is higher than the demand price at infinitesimally small quantities, and where at zero price supply exceeds demand. The latter is the definition of a free good, and these are not traded in markets. Within the range of potentially feasible products, that is those with a positive equilibrium price, only a sample of products can actually be observed. One question is why these products are on the market and not some other products, or some additional products. The economics literature lacks any formal answer to this question, and concentrates instead on the production and distribution of available products. Yet why a particular product is on the market and not some other version of it calls for explanation. The economic theory of clusters is a part of the explanation of why certain items are on offer and not others.

Economists are interested in non-market modes of economic organisation, such as government provision of public goods, and what goes on inside giant firms. But the market as an example of a self-organising system is a major concern. Along with the study of the characteristics of markets goes a concern with potential market failure. Clusters might be an example of market failure. Depending on the forces that bring about clusters, it might be possible to produce more with the same resources, or the same amount with fewer resources, or more profitably. So there is an efficiency concern and a welfare concern with the phenomena of economic clusters. In certain circumstances there could be scope for policy intervention with the goal of promoting or reducing the extent of clusters of a particular kind in particular circumstances. In general, Hotelling clustering is likely to be related to a kind of market failure. Information based clustering is likely to be socially desirable.

Economic analysis of clusters proceeds by drawing on information as to the nature or kind on the cluster in question. Is it a cluster in product space, in physical space or in some other dimension? Wholesale markets for vegetables, fish and meat are familiar clusters. In many major cities, these clusters have moved from central locations to closer to the perimeter of cities. Why do these clusters change location while many others do not? The kind of cluster it is can be a part of the explanation. Transport costs and land values may also be significant factors. Having pieced together an explanation, the question remains, is this a fairly accurate explanation? Several avenues of enquiry are available. Are there similar clusters in all relevant respects including land values and transport costs who did not move? Perhaps there is another significant factor, such as negative externalities, which has to be brought into the analysis.

This process involves the interaction of several theoretical and empirical elements. Many of the theoretical elements can be examined in other contexts. Relevance in related activities supports the assumption that it is relevant in this context. Frequently the explanation under consideration can be applied to other cases. Does it appear to give a plausible answer in those cases? Sometimes international comparison can greatly expand the number of relevant observations.

Which theory, or combination of theories, is to play a role in a particular investigation is usually determined by observations which are independent of the theories themselves. Patients on Harley Street are not usually looking for a doctor who is nearest to where they are. This immediately rules out Hotteling clustering. Economists tend to find out about what is going on through a process of investigative intuition. A variety of theoretical probes combined with factual considerations generate a candidate explanation. The theoretical element is employed with some confidence not because it is a 'law', but rather because of its possible relevance in a particular context. The process of economic enquiry of this kind involves continual movement back and forth between candidate observations and candidate explanations.

In an earlier period, the general view of economists was that all economic clustering was Hotelling in nature.

> This is a principle of the utmost generality. It explains why all the dime stores are usually clustered together, often next door to each other, why certain towns attract large numbers of firms of one kind, why an industry, such as the garment industry, will concentrate in one quarter of a city. It is a principle which can be carried over into other "differences" than spatial differences. The general rule for any new manufacturer coming into an industry is "make your product as much like the existing products as you can without destroying the differences." It explains why all automobiles are so much alike and why no manufacturer makes a car in which a tall hat can be worn comfortably. It even explains why Methodists, Baptists, and even Quakers are so much alike, and tend to get even more alike. (Boulding 1966, 484.)

In current thinking, economists would agree that some clustering is Hotelling in nature, but other clusters have other explanations. In the above quote, both the dime stores and the garment industry examples are wrongly explained by Hotelling. The cluster of retail outlets generally has more to do with information issues than transport costs. The garment industry example falls more in the domain of production externalities. Having laid out a brief history of economic analysis of clustering we turn to a two examples of cluster analysis as illustrations of theories in context.

7.4 Clusters in Heterogeneous Product Outlets

The paper by Fischer and Harrington (1996) provides a good example of the economic method of theories in context. The goal of the paper is to account for a particular example of clustering of retail outlets. Clustering can be observed in some products and not in others. The evidence for their paper comes from Baltimore in the United States in 1992. There is no immediate policy implication of the enquiry. The objective is purely to provide a reasonably accurate account of an economic phenomenon. Such knowledge could provide relevant information for economic policy in the future. For example, much building location and the use of buildings requires planning permission. Urban planning policy in Baltimore might benefit from a better understanding of the factors leading to this example of clustering. But the immediate objective is purely an understanding of the cause of this case of clustering of retail outlets.

The enquiry begins with an observation. We notice different degrees of clustering in different types of retail outlets, depending on the kind of products being sold. As we are concentrating on a very particular example of clustering, the analysis could be described as being *ad hoc*. This is often taken to be a pejorative term. Our view is that this is a mistake. The method of theories in context is inherently case specific, and there is nothing wrong with that. For certain investigations it is best and entirely appropriate to be *ad hoc,* and it is better to acknowledge that.

As indicated above, there are a range of theories of clustering. Which theory, or combination of theories, might account for what we observe? A very reasonable starting point is the conjecture that retail outlets want to appeal customers. The range of strategies to achieve this end might include advertising, pricing, and range of products on offer. This study concentrates on the location decision as a means of encouraging sales. Quite a number of factors might be involved in location: convenience, travel time, knowing where to go. Looking at the different kinds of retail outlets in Baltimore in this period, it stands out that clustering is stronger for some types of products. For antiques we see stronger clustering, and for some kinds of reasonably expensive electrical goods. What do the products of the clustered retail outlets have in common? The authors seize on the aspect of heterogeneity. The products are not simple and uniform. Consumers have different preferences with respect to these kinds of products. They also need to see the products and get additional information on the products on offer. So far we are dealing with guesswork and intuition. The next step is to ask, what does heterogeneity have to do with clustering?

To answer this question, we turn to theory. It is important to take note of a movement back and forth between observations and theory. The process is far from a one-way relation between a phenomena and an explanation. At this stage we can see that some aspects of the theory tentatively employed are received, or off-the-shelf, theory. Other aspects are specific to this enquiry. As we are dealing with consumer behaviour and retailer behaviour, natural starting points are utility maximising for consumers and profit maximising for retailers, as general propositions. But what are the relevant factors for this enquiry? Clearly consumers have to know where to search, and retailers have to decide where to locate. At this point the authors develop a model. Given the somewhat confusing state of the methodology literature on models, it is worth specifying that for our purposes, and those of Fischer and Harrington, a model is a specific and explicit statement of a more general theory. A number of considerations contribute to the need for modelling at this stage.

It is one thing to intuitively and casually suggest the heterogeneity of products, in a price range, leads to clustering. The suggestion may appear to make sense. But appearances can be quite misleading. One purpose of the model is to set out a fully explicit example relating heterogeneity and clustering. At this stage in the enquiry it is not crucial that the explicit model constructed be a perfect, or even a pretty good, description of what was going on in Baltimore in 1992. Of course, it would be nice if that was the case. But initially we have more modest need for the model. That need for the model is to answer the question can we conceive of *any* plausible mechanism that accounts for clustering through heterogeneity? Given the facility of economists with modelling, if we cannot construct a reasonable model that gives such a result, that failure would seriously put in doubt the suggested proposition. The desire for a 'reasonable' model is best illustrated by examining the model set out by Fischer and Harrington.

Some background for the model is helpful. As indicated, the goal of the exercise is to account for, or explain, the presence of certain retail outlet clusters. We want to know why they occur. They are not taken to be the only form of outlet location. Some firms locate outside the cluster, or in the 'periphery' in the language of Fischer and Harrington. It is cheaper for a consumer to search several firms in a cluster once a consumer has gone there. The location of the cluster means that some consumers find it cheaper to travel to the cluster than others do. There is great consumer uncertainty mainly because the product class is not uniform. Consumers differ in their preferences for product characteristics, and they do not know what is on offer from any outlet.

For Fischer and Harrington, the key factor is lack of knowledge of the nature of some products prior to inspection. Prices in the cluster are likely to be lower than in the periphery, though this is not necessarily the case with more extreme product heterogeneity. Given the likely advantages, why do not all consumers purchase from the cluster if one exists? There are two reasons. The cluster may be far away for some consumers raising travel costs, and not all varieties of products are available in the cluster. Even if a consumer has visited the cluster, the consumer may not purchase there and may continue looking in the periphery for the kinds of product he or she prefers.

The explicit model with propositions regarding firm and consumer behaviour uses the degree of heterogeneity as a key consideration. For most parameter values, the greater the heterogeneity, the greater the likelihood of a cluster and the higher the proportion of outlets in the cluster. This relation is not imposed by the model. It is derived from other assumptions. Under some circumstances, greater heterogeneity could result in more periphery firms due to the pricing effects. With finite choice of product variety in the cluster, greater heterogeneity can lead to more consumers searching in the periphery. But on balance, for most parameter combinations, the greater the heterogeneity the stronger the tendency to cluster. Other effects of greater heterogeneity are increased monopoly power in the periphery and reduced price competition in the cluster. In principle the net effect on prices of the relative size of the cluster and the periphery can go either way.

These considerations suggest that there is scope for considerable complexity in this type of clustering. Fischer and Harrington adopt an explicit model of consumer and retailer behaviour. Each firm sells a single variety of the product which is randomly assigned to the firm from a specified distribution. Firms face a fixed cost of entry and sell at zero cost. They choose a price intended to maximise profits. Firms are aware of the prices charged by other retailers. They know the number of firms in the cluster and in the periphery. Profits depend on the price chosen and on the location.

The unrealism of these aspects of this part of the model are interesting. The goal is that they should not bias the conclusion. Greater complexity would follow if the firms chose several products and had imperfect knowledge of the prices charged by other retailers. We could have theories about how they form views of pricing in the market. Indeed, some investigations address that issue as a central concern. Another possibility is that selling costs could be positive and vary with the amount sold.

Here we assume otherwise. Do these specifications in this exercise, which have the advantage of leading to a tractable model, significantly influence the results?

A central proposition for our view of theories in context is that there is no general rule about unrealistic assumptions or elements of the model. They may be entirely appropriate, or vey inappropriate. It all depends. It depends in part on the objective of the investigation. It also depends on whether the assumption is comparable to assuming the result or is a reasonable simplification which only yields the result in a plausible setting of other elements.

An important contribution of this modelling exercise is to indicate that a simple relation between variety of product types and the existence of a cluster is not the case. The observations and the intuition they suggest are informed and enriched through the modelling exercise. The model points to, or reveals, the possible importance of factors that the broad theoretical proposition does not consider. A lot hinges on whether greater realism in the propositions of the model is unlikely or likely to alter the conclusions. Of course, if another model, and particularly a more realistic one, generated a simple relationship less dependent on parameter values, that would be telling. It is always possible for another model to give a different slant. But what is in principle possible is very different from an actual demonstration.

Similar modelling strategies are adopted by Fischer and Harrington for incorporating consumer behaviour in this market. Consumers differ one from another in two ways. They have different rankings of product attributes. In other words, their preferences differ. They also differ in the cost incurred in travelling to the cluster. For the first consideration the authors assume a finite range of valuations of products. The valuation each consumer places on each product is assigned from a random draw from a uniform distribution. Consumers are uncertain about the value to them of the products on offer from different outlets until they arrive at the store. There is a common cost to all consumers for each search made in the periphery. The cost of searching in the cluster is assigned to each consumer according to a draw from a distribution. There is no cost to searching all the firms in the cluster after meeting the initial cost of getting to the cluster. Consumers are assumed to have an accurate knowledge of the number of retailers in the cluster, and to assume an infinite number of outlets in the periphery.

This last element is quite common in modelling and is not realistic. This has no negative effect on the analysis. What it means is that from the consumer point of view, there is an endless supply of firms in the periphery. As a practical matter, individuals can go on searching there without coming to an end in outlets. Consumers retain a memory of what they found in a search. They have the option of stopping searching, continuing in the periphery, or looking in the cluster. Consumers also have the option of stopping searching and not making any purchase. The market size is determined by the number of consumers. Each consumer makes one or zero purchases. Again, this modelling device has no substantive consequences. Multiple purchases could be thought of as more individuals making a single purchase.

Having specified in their model how retailers and consumers behave, the next step is to bring these elements together in the form of a market structure which in this case concentrates on a location structure. The authors search for a kind of equilibrium. Of course, this is unrealistic. There never is any settling down. Consumers come and go, new products appear, and so on. However, firms cannot change their location in response to every shift in the market. And with a large number of consumers, individual arrivals and departures may not matter very much. So, the application of the concept of an equilibrium as an outcome of interacting forces can validly inform. Essentially, the authors use a Nash equilibrium concept. In the equilibrium configuration of retailers in the cluster and in the periphery, no retailer has an incentive to change location. Consumers similarly have no incentive, as a group outcome, to change their shopping searches strategy.

For a wide range of parameter values the model predicts the existence of firms in a cluster, and some firms in the periphery. But this does not apply to all reasonable parameter values. It is possible in the model that there will be no cluster. If this is a reasonably accurate theory, one must conclude that for Baltimore in 1992 the relevant parameter values were such that clusters in certain heterogeneous products would occur. The Lipsey and Eaton (1982) model of comparison shopping is less complex, and interestingly it concludes that a cluster must occur.

Through the history of analysis of clusters we can see a great variety of explanations of particular kinds of clusters. Some studies, such as Fischer and Harrington, take as their starting point a particular observed cluster. Other studies, such as Lipsey and Eaton (1982), postulate a kind of cluster due to a specified information consideration and then spell out a model than explains that kind of cluster. A specific and explicit specification of the general proposition is on offer. And second, the model draws attention to parts of the market mechanism that lead to the result, or lack of a result, in a cluster. It suggests relevant considerations.

The fact that we have a plausible, or possibly explicit explanation of the presence of a cluster is a rather weak indication that we have the right explanation. For example, we may explain a cluster as being due to quality signalling. In cases where consumers are in a poor position to judge the quality of the service they are receiving, and quality is an important consideration, we may get a cluster of medical practitioners, for example. We can specify a model where patients gravitate to the cluster, and especially able doctors choose to practice there. The realism of assumptions is of some consideration, but even if this test is met, do we have the right model? The model may fall within the bounds of the general proposition, i.e., quality signalling can cause a cluster, but the general proposition may still be wrong. Much better would be if the model had other implications in addition to the one it was designed to imply, and we could examine the consistency of these implications with evidence. This is rare in economics. Hopefully, as the discipline evolves it will become more common.

A general discussion of the role of unrealistic assumptions in economic modelling is of some value, but of limited value. The example above, and the one that follows, highlights the sensitivity of modelling to the goal of the exercise. The same assumption, clearly unrealistic, may be appropriate, or at least innocuous, in one context and very misleading for the analysis in another.

7.5 Clusters in Television Programming

We conclude with an example of empirical work using an Hotelling explanation of clustering. The paper we have chosen is a policy focused investigation. Peter Steiner did extensive work on the competitive efficiency in radio broadcasting (cf. Steiner 1952). As with any normative exercise, the goals of policy are largely exogenous to economic analysis as such.

> Every enquiry into the workability of competition rests upon social judgements made with respect to some criteria of public interest; as such it becomes a problem in what is generally designated welfare economics, and the selection of appropriate criteria becomes particularly important. The economist must appraise the adequacy of means to ends, and he must also attempt to define the appropriate ends, though in doing so he is placed in a rather more exposed position than he has traditionally chosen to occupy. (Steiner 1952, 194.)

Some years after his initial interest in radio broadcasting Steiner turned his attention to the case of television in the United Kingdom in 1961 (cf. Steiner 1961). At that time there were two channels broadcasting, the public service channels of the BBC and the ITV commercial channel supported by advertising. The social judgement Steiner chose to make was informally formulated and amounted to taking as a goal that the offerings should satisfy the largest number of viewers. This goal is difficult to make operational. Much easier to apply is the test of avoiding duplication. If two or more channels put out essentially the same programme, there is a strong *a priori* case for an unnecessary loss in welfare. A broadcast is a public good in the sense that viewers are not excluded from watching and any number can watch without diminishing the experience of other viewers. Rather than duplicating, any reasonably different offering would add to welfare as at least some viewers would have viewing closer to their preferred programme.

Steiner made careful note of United Kingdom television broadcasting in 1961 over one Sunday broadcasting and over the week 1st to 7th January. The major observation is of considerable duplication, or close similarity of offering. There is some judgement involved in what constitutes duplication as no two programmes are exactly the same. However, he found a case where, for example, the two channels were putting on original plays at essentially the same time. It was common practice for one to start a few minutes before the other in an obvious effort to steal the audience. Two blockbuster 'specials' involving a number of celebrity participants were frequently scheduled close to each other. Steiner gave some attention to the question of who had access to the scheduling commitments of rival channels in making their decisions on scheduling.

This example fits very nicely into an Hotelling theory of clustering. The clusters are in product space where we observe similar, or near similar offerings on the two channels. Clearly this is not information driven clustering. Consumers can virtually costlessly look up what will be on offer or failing that can simply see what they are getting. The Hotelling assumption for viewers is that they will watch the programme closest to their preference. The assumption for channel providers is that they will attempt to maximise their audiences.

We can characterise the problem with a hypothetical example of an audience where 80 % of the viewers want to watch a western and 20 % prefer a ballet. This is the television equivalent of the ice-cream sellers on a beach, the difference being that on the beach the consumers are spread evenly, where in this TV example they fall into two groupings. On those assumptions, a single channel would get all the viewers whatever it put on, but might kindly offer a western, ignoring costs of production. Moving from monopoly to two channels, both being audience maximisers they both will put on a western. The same for three and even four channels. Westerns proliferate. With five channels every scheduler would be indifferent to putting on a western or a ballet, as long as there was the only one ballet. As the number of channels goes up, the Hotelling model predicts both clusters, and channels located in the 'periphery'. The key policy point is that a cluster on these assumptions reduces welfare. The viewing maximising objective means the public could be better served with no duplication and more variety. Steiner's empirical investigation of United Kingdom television implies that the Hotelling effect is operative.

Hotelling clustering is simpler than other types of clustering, such as information driven clustering. The reason for this is that only the producers determine the equilibrium. In more complicated models where the consumers respond to the channels on offer by, for example, choosing not to view, a more complex equilibrium condition is required. The analysis has important implications for public service broadcasting. Ideally the public service channel would not act like a commercial channel. It would not try to maximise its audience. Rather, it would look at the over-all offering and fill in existing gaps. In the western/ballet example, even in the two channels case it would put on a ballet which would not maximise its audience, but would improve welfare. The Hotelling analysis provides a strong policy prescription based on a theory of economic behaviour combined with a welfare objective.[3] The result depends on the application of the theory in a particular context.

The theory does imply that as the number of channels gores up more and more of the public will have their preferences catered to, along with some local clustering. In the United Kingdom today we can see both the BBC and ITV offering news at ten. But more disturbing for the theory is the common observation, if correct, that even with on the order of six hundred channels a large section of the public often cannot find something to view. This assertion calls for some serious investigation, but pending that, there is a common feeling that in spite of the large number of channels, television today does not cater well for the more discerning and perhaps better educated sector of the potential audience. Perhaps the theory that worked quite well in one context is significantly less applicable in another context.

---

[3]The BBC does not seem to adopt this policy recommendation. This could be due to a lack of understanding of the Hotelling implications for TV on the part of the BBC and of Parliament. Perhaps the licensing fee system may lead to the view that as all viewers pay, the BBC should try to provide for the largest number, making it very much like a commercial station.

7.6   Conclusion

Our claim is that economic clustering is a useful case study of the way in which much of economics works. A key point for us is that this example of empirical investigation of an economic phenomenon does not rely on laws of economics. Instead, tentative explanations are related to observations. The process of enquiry evolves and develops. Over time there is back and forth movement. Theory leads to new observations, and the evidence in turn leads to new theories. All very well, one might say, but is this science, or just common sense? Many activities, such as planning a holiday, can benefit from informed investigation. Past experience comes into it as well as conjectures about costs and benefits of choosing different seasons for the holiday. It is a rational activity, but it is not science.[4]

A deep investigation of the philosophy of common sense would take us too far afield, and hopefully is not required for our purposes. Our goal is to argue the case for economic science without laws. Our claim is both that this is how a large section of economic activity in fact takes place, and that this is appropriate for this social science. We further feel that to dismiss the notion of theories in context, which is another way of saying theories which *are* more than a little *ad hoc*, as being simply common sense is misleading. There are several reasons for this. One has to do with the cumulative feature of economic understanding.

For many investigations, ideas about how participants in a particular economic arena behave are central to the explanation. (Some would claim they are essential, or at least highly desirable, in all economic investigations. This is the micro foundations school of thought. We need not enter into that debate here.) That certainly is the case with the theory of clustering, or at least for the examples we have chosen to highlight. Economics is famous, if not notorious, for the use of extremely simple notions of human behaviour. This criticism is wide of the mark. In fact, economists are open to richer and more complex theories of how people respond in specified situations, provided that they inform the analysis. This is different from asking for richer theories of behaviour because they are more realistic. Over the past 80 years or so, economics has undergone enormous changes in approach through employing different assumptions about behaviour. The theory of clustering is a good example of some changes. Much of the early work assumed that both consumers and suppliers, or producers, were in possession of all the relevant information they needed in order to decide on a course of action. Later developments in clustering theory allowed for a lack of different kinds of information on the part of consumers; information about prices, outlet locations, and quality of offerings, for example. This cumulative nature of economic enquiry is one aspect which distinguishes it from common sense.

---

[4]Lewis Walport maintains that science is "unnatural," meaning that it is not common sense (cf. 1992). On that reading, if 'theories in context' is simply common sense then what we are describing is not science. We suggest that this aspect of economic method is more than common sense.

The remarkable growth and change in economic theory over recent decades along with greater facility in modelling and in huge strides in statistical inference suggests there is progress in the discipline.[5] But this does not guarantee better understanding of future economic events or the discovery of better policy prescriptions for dealing with future events. The simple reason for this is that the economic world is changing along with the changes in human society generally. The complexity of events and the impossibility of experimentation on an adequate scale means that even our understanding of well-studied past events is imperfect. When it comes to the future, the potentially changed nature of society means that development in the discipline will not ensure continually better performance of it over time. Hopefully, on many occasions there will be better performance, but we cannot count on that generally. The lack of laws of economics places the discipline in this unenviable position.

Common sense and intuition appear to be closely allied. The application of theories in context is far from a reliance on intuition. Care with well-specified models along with care with observations takes this branch of economics well away from common sense. We believe we have described the working method of a large branch of economic enquiry in an accurate manner. We do not claim that theories in context is the only possible approach to the understanding of economic phenomena. Everything has to be tried. There are largely historical approaches, simulation exercises, and many others. We are not proposing a methodology of all empirical investigation in economics. We are setting out a view of the method of one very large class of economic enquiry.

Bibliography

Asarni, Y., & Isard, W. (1989). Imperfect information and optimal sampling in location theory: An initial re-examination of Hotelling, Weber and Von Thunen. *Journal of Regional Science, 29*(4), 507–521.

Borenstein, S., & Netz, J. (1999). Why do all the flights leave at 8 Am?: Competition and departure-time differentiation in airline markets. *International Journal of Industrial Organization, 17*(6), 611–640.

Boulding, K. (1966). *Economic analysis, volume 1: Microeconomics* (4th ed.). New York: Harper and Row.

Brown, R. (1984). *The nature of social laws.* Cambridge: Cambridge University Press.

Christou, C., & Vettas, N. (2005). Location choice under quality uncertainty. *Mathematical Social Science, 50*, 268–278.

Downs, A. (1957). *An economic theory of democracy.* New York: Harper and Row.

Dray, W. (1957). *Laws and explanation in history.* Oxford: Oxford University Press.

---

[5]Remarkable examples are: strategic analysis, much of it in the form of game theory; behavioural economics with less than perfect reasoning; institutional economics, including what goes on inside firms; the information revolution and asymmetric information; macroeconomics released from the textual grip if Keynesianism, and many others. Significantly, these changes have all come from the mainstream technical part of economics, not from the disaffected critics.

Fischer, J. H., & Harrington, J. E. (1996). Product variety and firm agglomeration. *The Rand Journal of Economics, 27*(2), 281–309.

Friedman, M. ([1953] 1966). The methodology of positive economics. Reprinted in: M. Friedman. *Essays in positive economics* (pp. 3–43). Chicago: University of Chicago Press.

Hausman, D. M. (1992). *The inexact and separate science of economics.* Cambridge: Cambridge University Press.

Hotelling, H. (1929). Stability in competition. *Economic Journal, 39*(153), 41–57.

Lipsey, B. C., & Eaton, R. G. (1982). An economic theory of central places. *The Economic Journal, 92*(365), 56–72.

Little, D. (1993). On the scope and limitations of generalizations in the social sciences. *Synthese, 97*(2), 183–207.

Morgan, M. (2010). *The world in the model: How economists work and think.* Cambridge: Cambridge University Press.

Osborne, M. J. (1993). Candidate positioning and entry in a political competition. *Games and Economic Behaviour, 5,* 133–151.

Robbins, L. (1935). *The nature and significance of economic science* (2nd ed.). London: Macmillan.

Steiner, P. (1952). Programme patterns and preferences and the workability of competition in radio broadcasting. *Quarterly Journal of Economics, 66*(2), 194–223.

Steiner, P. (1961). Monopoly and competition in television: Some policy issues. *The Manchester School of Economic and Social Studies, 39*(2), 107–131.

Wolinsky, A. (1983). Retail trade concentration due to consumer imperfect information. *Bell Journal of Economics, 14*(1), 275–282.

Part IV
Methodological Pluralism in Natural Sciences and Sciences of the Artificial

Chapter 8
Plurality of Explanatory Strategies in Biology: Mechanisms and Networks

Alvaro Moreno and Javier Suárez

Abstract Recent research in philosophy of science has shown that scientists rely on a plurality of strategies to develop successful explanations of different types of phenomena. In the case of biology, most of these strategies go far beyond the traditional and reductionistic models of scientific explanation that have proven so successful in the fundamental sciences. Concretely, in the last two decades, philosophers of science have discovered the existence of at least two different types of scientific explanation at work in the biological sciences, namely: mechanistic and structural explanations. Despite the growing evidence about the radically different nature of these two types of explanation, no inquiry has been conducted to date to determine the ontological reasons that might underlie these differences, nor the way in which these types of explanations can be systematically related with each other. Here, we aim to cover this gap by connecting this plurality of research strategies with the existence of emergent levels of reality. We argue that the existence of these different—and apparently incompatible—explanatory strategies to account for biological phenomena derives from the existence of "ontological jumps" in nature, which generate different regimes of causation that in turn demand the development of different explanatory frameworks. We identify two of these strategies—mechanistic modelling and network modelling—and connect them to the existence of two ontological regimes of causation. Finally, we relate them with each other in a systematic way. In this vein, our paper provides an ontological justification for the plurality of explanatory strategies that we see in the life sciences.

Keywords Scientific explanation · Mechanism · Structural explanation · Networks · Emergence

A. Moreno (✉)
Department of Logic and Philosophy of Science, IAS-Research Centre for Life, Mind and Society, University of the Basque Country (UPV/EHU), Donostia-San Sebastián, Spain
e-mail: alvaro.moreno@ehu.eus

J. Suárez
Department of Philosophy, University of Bielefeld, Bielefeld, Germany
e-mail: javier.suarez@uni-bielefeld.de

© Springer Nature Switzerland AG 2020   141
W. J. Gonzalez (ed.), *Methodological Prospects for Scientific Research*, Synthese
Library 430, https://doi.org/10.1007/978-3-030-52500-2_8

8.1   The Ontological Basis of Explanatory Plurality

In his book *Mind and Cosmos. Why the Materialist Neo-Darwinian Conception of Nature is Almost Certainly False* (2012), the American philosopher T. Nagel argues that contemporary biological sciences are unable to explain neither the origin of life, nor the origin of mind. More fundamentally, Nagel claims that no scientific theory can provide a satisfactory account—which for him means a reductionistic explanation in physicalist terms—of how these two complex phenomena could have appeared from the evolution of the physico-chemical world. In his book, Nagel tends to identify the epistemological or explanatory sense of the term "reductionism" with the ontological one, i.e. he conceives "reductionism" as equivalent to "reductive materialism", as he does not consider that emergentist theories could offer an alternative explanation of the origin of these types of complex systems (and their related properties) in ways compatible with materialism. As he says (Nagel 2012, 55–56): "That such purely physical elements, when combined in a certain way, should necessarily produce a state of the whole that is not constituted out of the properties and relations of the physical parts still seems like magic even if the higher-order psychological dependencies are quite systematic."

We take Nagel's example to introduce here why a correct understanding of "the ontological jumps" in nature is so important, and why it is fundamental to properly characterize and conceive the different explanatory strategies that are currently being employed in the life sciences. As it is well known, emergentism—the theory that aims to account for the existence of these ontological "jumps"—appeared in the 1920s as an attempt to find a compromise between two theses: on the one hand, the fidelity towards a basic materialism; on the other, the assumption of the appearance of novel levels of reality (chemical bonding, living systems, the mind) endowed with specific causal powers (cf. Mill 1843; Alexander 1920; Lloyd Morgan 1923; Broad 1925). Of course, the crux of emergentist theories has always been the difficulty to provide a scientifically satisfactory account of how certain complex arrangements of matter can bring forth new interesting properties, and why the appearance of these new properties cannot be reduced to—i.e. why they cannot be logically deduced, at least in principle, from—the theory that works successfully for the explanation of the properties of the lower-level arrangements of matter. In other words, why certain complex relations suddenly generate what are conventionally called "systemic" properties, and why these systemic properties should be qualified as ontologically "genuine" and different from the properties of the lower levels of reality (i.e. emergent), rather than "subjectively surprising".

Emergentism experienced an important decline since the 1930s, as it remained suspicious both for most scientists and philosophers, especially due to the influence of logical positivism (Kim 1989, 1999, 2006; McLaughlin 1992, 1997; Humphreys 1997; Wilson 2013, 2016; Suárez and Triviño 2019). There are many reasons that justified this suspicion—if not open rejection. Firstly, in agreement with the main claims defended by logical positivism, reductionism was perceived as the only possible way to scientifically account for the properties of "higher" or more complex levels of the reality. Secondly, if—as overwhelmingly new discoveries were showing—higher levels of reality (like biological systems) were made *but of* physical entities, their properties should, at least "in principle", be explained in the very same terms of the physical sciences. Finally, the possibility of reducing Classical Thermodynamics to Statistical Mechanics constituted an important ontological victory for the reductionists. It showed how a set of laws that had been previously considered emergent could in the end be reduced to the laws of physics, and thus their alleged ontologically distinct explanatory framework finally converged into the explanatory form of the so-called "fundamental" (namely, the physical) sciences.

Nonetheless, the decline of the influence of logical positivism among philosophers of science, together with the development of the so-called "Sciences of Complexity", have generated a resurgence of emergentism since the last decade of the twentieth century, both among philosophers and scientists (Witherington 2011; Mitchell 2012). In parallel with this resurgence of emergentism—and partially as a consequence of the growth of the Sciences of Complexity as well—philosophers of science have started to admit the existence of a plurality of forms of scientific explanation which go far beyond the type of reductionism that Nagel argues to be necessary in his book (e.g. Dupré 2013; Díez et al. 2013; Rice 2015; Green and Jones 2016; Green 2017; Woodward 2017). In the specific case of the life sciences, two alternative forms of scientific explanation have gained special recognition in the last two decades: mechanistic and structural explanations (Huneman 2010, 2018a, b; Brigandt et al. 2017; Suárez and Deulofeu 2019). The acknowledgment of this plurality of explanatory strategies has given rise to a blossoming of research that is raising very interesting questions, and that is questioning some of the most traditional assumptions about the nature of scientific explanation, as we will show later.

Despite the complementary aspects of these debates, these two dimensions of our current philosophical theorizing have not been connected in a consistent manner yet. Philosophers of science have simply confirmed—by providing several elaborated examples to that end—the existence of a plurality of explanatory strategies which seem to have a radically different nature from each other, but they have not conducted any inquiry about the ontological reasons that are behind this plurality. This chapter aims to contribute to this growing literature by covering this specific gap. Here, we raise the question about how the ontological and the epistemological debates can be systematically related by showing that the different explanatory forms that we observe in the life sciences are a consequence of the existence of "ontological jumps" in nature. These "jumps", we argue, generate emergent regimes of causation that, first, constrain the application of the laws of nature in a particular domain of reality and, second, justify the appearance of these apparently incompatible explanatory frameworks for each of these domains. Importantly, our study will be focused on the biological sciences. We will inquiry first about the nature of the type of explanations that work well in this domain; second, we will ask about how and why successful explanations in the biological sciences are related to successful explanations in more fundamental sciences.

To do so, the chapter is structured as follows. In Sect. 8.2, we introduce three different types of explanatory strategies that have been used in different sciences—concretely, in different parts of the biological sciences—and show why they are epistemologically incompatible. In Sect. 8.3, we investigate the ontological reasons that underlie the validity of mechanistic explanations, and we explain their relation to successful explanations in the fundamental sciences. In Sect. 8.4, we conduct the same type of investigation for structural explanations. In Sect. 8.5, we argue that, in view of the analysis conducted in the previous sections, biology requires complex explanations, and thus both the mechanistic and the structural explanatory strategies are necessary and need to be sometimes combined. Finally, in Sect. 8.6 we present our conclusions.

8.2   Three Types of Explanatory Strategies: Deductive-Nomological, Mechanistic, and Structural

As we said before, in his book, Nagel (2012) presumes that the only way of providing a satisfactory scientific explanation of a phenomenon is to do so in physico-chemical terms, as these are the terms employed in the so-called fundamental sciences. Admittedly, the type of scientific explanation that the fundamental sciences have so successfully developed implicitly requires three elements: First, the identification of the relevant observables (state variables); second, the determination of the general laws governing their change (expressed as differential equations, state transition rules, maximization/minimization principles, etc.); and third, the determination of the initial (or boundary) conditions for each particular case. Once these three elements are given, it is possible to provide an explanation of why several physico-chemical phenomena happen, as well as why they occur in the way they do, rather than differently. For example, it is possible to explain why an ice cube melts into water by appealing to observables such as the temperature, pressure, or internal energy of the ice cube, plus a law of nature that says that ice melts whenever it is left at a temperature about 0 °C, plus a set of initial conditions of the case—e.g. that the ice cube is originally at a temperature of −15 °C. This type of explanation has been called *deductive-nomological* (DN, hereafter) insofar as in this type of explanation the phenomenon to be explained is the conclusion of a logical argument whose premises must necessarily include—explicitly or elliptically—*a law of nature* (Hempel 1965; Díez 2014; Deulofeu and Suárez 2018). The law of nature describes the regularity at the fundamental level of reality whose existence serves to predict the phenomena that will be observed and, thus, explains their occurrence. In this vein, the DN approach to scientific explanation conceives explanations in analogy with predictions: a prediction tells you what will happen in the future by appealing to the regularity imposed by the laws of nature, whereas an explanation tells you what has occurred in one particular instance in the past by appealing to the very same regularity.

DN strategies, despite their application in some parts of the physico-chemical sciences, cannot work for explaining the global properties of systems made of an enormous number of parts or elements, though. This difficulty has been however overcome in some cases, such as Statistical Mechanics, where it is possible to average the huge amount of observables and thus reduce the explanations in a different field—Classical Thermodynamics—to explanations in terms of the more fundamental science. However, there are many other systems in which the interacting units cannot be averaged out and, in some cases, they even interact in non-linear and highly selective ways. Actually, this is what happens in those systems that are conventionally called "complex" (such as living systems), and that are the object of study of the Sciences of Complexity.

Some philosophers of science have argued that the complexity of this type of systems makes impossible that any fundamental law of nature could capture their properties and behavior to make them compatible with the strict requirements that DN explanatory strategies pose (Mitchell 1997, 2003; Giere 1999; Woodward 2000). Yet, despite this huge complexity, complex systems often exhibit global (or "emergent") simplified properties. And the causal connection between the messy set of interactions of the components and the "emergent" global properties remained unexplained. These systems have challenged scientific understanding because their behavior could not be predicted, and their properties were never fully explained through available mathematical models, based on the fundamental laws of nature (Moreno et al. 2011).

And yet, despite their overwhelming complexity, living systems have been successfully studied for centuries, following a very different explanatory strategy than that of the physical sciences that inspired DN models of scientific explanation. Rather than looking for universally applicable laws or predictive models, biologists have tried to understand the behavior of living organisms by decomposing them into various parts ("structures"), analyzing them separately, and, then, investigating how these parts interrelate and affect one another within the whole system. Note that the point here is epistemological: the point is not denying that the laws of physics do not apply to these systems, but rather that these systems have a structure that makes these laws useless to explain their behaviour. This explanatory strategy—called "mechanistic"—has provided fundamental knowledge about the basic structure and about the integration of many functional parts of living systems.

The mechanistic strategy to study living systems has a long history (cf. Nicholson 2012, 2018), although it has only become popular in analytic philosophy of science in the last two decades, where it is usually characterized as "new-mechanism". New-mechanism was originally introduced in Bechtel and Richardson (1993) and Glennan (1996), and it acquired popularity after the publication of Machamer et al. (2000). New-mechanists share a basic commitment to the two following theses. First, the rejection to the claim that scientific explanations are arguments, as it was widely assumed among defenders of the DN approach; second, the belief that the analytic decomposition of a complex system into their more simple components permits that relatively few parts could be methodologically isolated from the rest such that causal mappings between specific functional operations and their distinguishable structural components could be drawn.[1] Combining these two theses, new-mechanists argue that to explain why a phenomenon occurs consists in providing a mechanism whose action is causally responsible for the production of the phenomenon to be explained. For instance, to explain why nervous signals travel from one neuron to another, it is necessary to explain how the mechanism of synapsis works, specifying how an electric signal can be transformed into a chemical signal (neurotransmitter), and then back into an electrical signal. What plays the explanatory role in this case, new-mechanists argue, is the specification of the mechanism itself, rather than a set of physico-chemical laws of nature, as defenders of the DN approach assumed.

In general, a mechanism can be defined as a set of entities—also called "parts"— and activities—also called "operations"—organized in such a way that the activities between the entities are causally productive of regular changes from start or set-up to finish or termination conditions (cf. Machamer et al. 2000; Glennan 2002; Bechtel and Abrahamsen 2005; Craver 2007; Craver and Darden 2013). In other words, a mechanism is an arrangement of parts (a structure) that performs a function because of the activities that the set of parts engages in as a consequence of the way they are organized. In this vein, a mechanistic explanation must always include two necessary and sufficient components to be satisfactory: first, the *model of the mechanism* (entities, activities, organization); second, a *causal story* which connects the orchestrated functioning of the model of the mechanism to the phenomena that are produced and that are required to be explained (Issad and Malaterre 2015, p. 270).

Given this characterization of mechanistic explanations, it becomes necessary to specify the criteria to single out the parts of the mechanism whose interaction causally produces the phenomena to be explained. In other words, as mechanisms are arrangements of parts ("structures") whose causal interactions perform a *function*—understood as a set of selected (see Sect. 8.5 and footnotes 4 and 5) terminations conditions—it is necessary to provide a criterion to distinguish the arrangement of parts that the mechanism comprises from other arrangements of parts. One possible way of solving this problem has been adopted by Craver (2007, p. 144), who appeals to Woodward's (2000, 2003) "difference-making" account of causation to single out the arrangements of parts that constitute the mechanism, *vis à vis* those elements that are considered external. In this account, to determine the arrangements of parts that constitute the "boundaries" of the mechanism, one must intervene in the parts of the mechanism, altering their behavior, and observing whether the alteration has any effect on the production

---

[1] It is important to note, although in passing, that the new-mechanist's commitment to the existence of *causal mappings* between a functional operation and the components that bring those operations about does not necessarily commit all new-mechanists to either to a physical reductionism (e.g., Salmon 1984), or to an "ontic" (e.g., Craver 2014) interpretation of mechanistic explanation. Rather, new-mechanists often assume a hierarchical structure of the world and believe that different mechanisms can be found out at different levels, realized by different entities, activities and forms of organization (cf. Krickel 2018).

of the phenomenon. This results in a view according to which "a part is causally relevant to the phenomenon produced by a causal mechanism if one can modify the production of this phenomenon by manipulating the behavior of the part, and one can modify the behavior of the part by manipulating the production of the phenomenon by the causal mechanism" (Nicholson 2012, p. 160).

New-mechanism has been the mainstream theory of scientific explanation at least since the publication of Machamer et al. (2000). However, in recent years, things have changed in philosophy of science. On the one hand, some philosophers have pointed out that the new-mechanistic explanatory strategy is not suitable to deal with several aspects of the biology of organisms, including some features of development, reproduction, or many other biological processes (cf. McManus 2012; Dupré 2013, 2017; Nicholson 2013, 2018; Issad and Malaterre 2015; Alleva et al. 2017). On the other hand, the new-mechanistic strategy seems simply not suitable to deal with many of the properties of "complex" systems that are explained by making use of the tools provided by the Sciences of Complexity (cf. Huneman 2010, 2018a, b; Jones 2014; Green and Jones 2016; Brigandt et al. 2017; Deulofeu et al. 2019). Concerning the later point, the application of the new-mechanistic strategy to account for the properties of complex systems would consist in assuming that complex systems result from the sum of several mechanisms working in a coordinated manner, so that the final outcome is the result of this combination, in the same way as it occurs in a classical mechanism. This type of explanatory strategy, new-mechanist argue, can be considered mechanistic as well, provided that the concept of "mechanism" is redefined accordingly (cf. Bechtel and Abrahamsen 2005; Bechtel 2015a, b). But, as we will argue, this explanatory strategy does not seem to fit well with the requirements of the new-mechanistic conception of explanation, and thus we are left with another explanatory strategy that following Huneman we call "structural strategies".

Structural explanatory strategies rely on the use of new techniques—such as cellular automata, genetic algorithms, Boolean networks, chaos and dynamical systems theory—derived from graph theory. These strategies, which we will globally call "network modelling", provide rigorous ways to study the emergent properties of a great variety of "complex" biological systems in a way that contrasts sharply with the new-mechanist strategy. While the key element that defines mechanistic explanations is the fact that the mechanism *produces* the phenomenon that is being explained, bringing a set of entities and activities from certain set up conditions to a termination condition, *no phenomenon is produced by any mechanism* in structural explanations. Rather, structural explanations work because the empirical system whose properties require explanation is identified with a mathematical structure whose properties can be mathematically studied. Structural explanation obtains in virtue of the relations of identity between the empirical system and the mathematical system, irrespectively of the causal mechanisms that bring about the properties in the empirical system. Thus, according to this view, structural explanations could be considered explanatory in Kitcher's (1989) sense of providing a "unification", because they (in a certain sense) connect a diversity of empirical systems subsuming them under a set of basic patterns and principles that are mathematical. Notice that the point here is merely epistemological, as no one would in principle deny that the final outcome that is produced in these systems results from the action of causes (see Sect. 8.5). The point that defenders of structural explanation have raised is that, whatever these causes are, they are not structured in a mechanistic way, and knowing them is irrelevant to explain the properties of the complex system—as the techniques used in the Sciences of Complexity demonstrate.

Take the example of the stability behavior of the microbiome, which has been systematically studied in Deulofeu et al. (2019). The stability behavior of the microbiome refers to the well-documented empirical observation that establishes that the species that compose the human microbiome keep their relative densities constant during the human's lifetime.[2] The stability behavior of the microbiome is counter-intuitive because the microbiome is affected by constant perturbations that, in principle, should alter this stability. Therefore, the fact that the microbiome shows this type of robustness requires an explanation. Deulofeu et al. demonstrate that the stability behavior of the microbiome is not explained in virtue of any mechanism that regulates the microbial species' densities, but in virtue of the network-like structure that is attributed to the system. They argue that the reason why the explanation of the stability behavior of the microbiome cannot be mechanistic is that, even though there seems to be a model of the mechanism defined in terms of the entities, activities, and their organization, all the details of the causal story that brings about the termination condition—stability behavior—are lost in the complexity of the mathematical analysis. As a mechanistic explanation necessarily requires the existence of a causal story, Deulofeu et al. argue that this explanation is non-mechanistic, but structural. If defenders of the view that all explanations—including structural explanations—are mechanistic want to redefine the concept of mechanism to fit these examples (e.g. Bechtel 2015a, b), they would have to renounce to the requirement of a causal story. But doing so seems to entail renouncing to the most basic ontological commitments of the concept of "mechanism", as we will explain in Sect. 8.4 (see also Issad and Malaterre 2015; Deulofeu et al. 2019).

In view of the previous example we can assume that a key element of structural explanation is that, rather than relying on the mechanism that causally produces a phenomenon, they rely on the mathematical (network-like) properties of the system. This is a fundamental element of the explanatory strategy that dominates the study of complex systems. Instead of trying to figure out how a phenomenon is causally produced, the structural strategy consists in uncovering the organizational features that generate an emergent property in a system, despite all the possible range of activities that can alter its mechanistic details. In that vein, structural explanations are non-causal—in the mechanistic sense of causality—but are still explanatory of some properties of complex systems, in virtue of the form of organization that these systems realize (Huneman 2018c). In other words, structural explanations rely on

---

[2]Notice that "species density" is an emergent property of the system, i.e., it only exists in so far as there is a microbiome. We will say more about this in the rest of the paper.

some sorts of mathematical properties of the empirical system—its topology, its tendency towards reaching a point of equilibrium, etc.—and abstract away from the causal details that the components of the system engage in and that lead to the realization of the property. The explanation is thus possible in virtue of this abstraction, plus the tendency of the system to realize the property in virtue of its mathematical structure alone.

This overview leaves us with three different and non-compatible types of explanatory strategies. Firstly, *DN strategies*, overwhelmingly used in the physical sciences and that rely on the existence of laws of nature to explain the phenomena in that realm. Secondly, *mechanistic strategies*, commonly used in many parts of the biological sciences—i.e., molecular biology—and that rely on the existence of arrangements of parts whose interaction causally produces a concrete outcome. Thirdly, *structural strategies*, employed in the sciences of complexity and that are based on the existence of a robust tendency in the system whose particular conditions are very hard to change.

8.3 The Ontology of Mechanistic Explanations in Biology: From Physico-chemical Laws to Mechanisms

As we have pointed out, mechanistic explanations have been—and still are—extremely successful in many parts of biology. But we have not explained the ontological reasons why this is so. To answer this question, we have to deal with two different problems. On the one hand, we should explain how systems that are ultimately governed by physico-chemical laws can coherently be explained by a form of mechanisms, which implies a form of explanation radically different from that of the DN explanatory strategy. On the other hand, the question is how these mechanistic explanations, which are typically reductionist and analytic, could be coherently related with the recent holistic structural approaches. We will try to answer the first question in this section, and we will develop the others in Sect. 8.4.

So, the question we are now asking is how a highly complex domain of lawful physico-chemical relations can give raise to another domain of relations where the units that interact are physico-chemical structures such that their interrelations are neither determined by physico-chemical laws, nor by systemic properties. For, as we explained, to talk about mechanisms requires implicitly assuming the existence of a world of units—mechanisms—capable of adopting a diverse and potentially indefinite number of relationships with one another in virtue of the nature of the entities that conform the unit, the type of activities they can engage in, and the way in which these entities are organized (cf. Machamer et al. 2000; Glennan 2002; Bechtel and Abrahamsen 2005; Craver and Darden 2013; Craver 2007; Arnellos and Moreno 2012). There are three key ideas underlying this conception of mechanisms (cf. Nicholson 2012; Militello and Moreno 2018).

Firstly, the idea of a stable organization of sets of *constraints* harnessing the action of physical (or chemical) laws. In fact, *what characterizes a mechanism are not the physico-chemical laws, but how the particular arrangement of selective constraints that the mechanism poses harness these laws* in such a way that the phenomena under study are systematically produced (Polanyi 1968). Or, to express it differently, how the model of the mechanism determines the causal story that in turn produces the phenomena that scientists aim to explain, in a way that does not directly depend on the action of the physico-chemical laws that ground the existence of the mechanism.

Secondly, the notion this set of constraints (the "structure" of the mechanism, or the model of the mechanism) serves a function or purpose. As the Webster dictionary defines it, a mechanism is "a process, technique, or system *for achieving a result*" (italic is ours). Namely, a mechanism is a set of constraints performing an "engineering-like" process.[3] For example, nobody would consider that a random set of rocks, which in a river speed up the flow of water by narrowing the bed, constitutes a mechanism because this effect is not functional. As we said in the definition of mechanism that we introduced in Sect. 8.2, the mechanism needs to produce the phenomenon, that is, it needs to lead the process from a set of initial conditions to the termination conditions.

Thirdly, the implicit assumption that through processes of arrangement of parts, an open[4] world of functions can be achieved. In other words, the idea is that mechanisms are the result of *compositional* processes of arrangement of parts, namely, the idea of construction.[5]

Interestingly, and connected to these three assumptions, we argue that what is behind the existence of a *mechanistic explanatory strategy* is a system constituted by a set of ontologically mechanistic relations. And behind a mechanistic relation what we find is *a set of constraints* operating on statistical thermodynamic flows. By constraint we mean a material structure that, in a given system, harnesses the (lawful) microscopic process (cf. Pattee 1972; Umerez and Mossio 2013) (Fig. 8.1). Suppose, for example, a set of chemical reactions. Chemical processes consist of molecular transformations (through chemical reactions) from a set of molecules to another different set. Chemical reactions encompass changes that only involve the positions of electrons, explained in terms of physical (quantum) laws. In addition to these laws, chemical processes are also governed by the second law of thermodynamics, which means that the final products of these chemical processes must be a set of highly stable equilibrium compounds. Thus, we argue that it is precisely the existence of this set of constrains one of the elements that makes

---

[3] However, this "engineering-like" process could also be achieved by a non-intelligent agent, like natural selection in non-human made systems.

[4] "Open" does not mean unlimited. The idea is that parts should be amenable to combinatorial processes, namely, that the combination of parts is not determined by intrinsic, lawful interactions.

[5] By this we mean a scenario where components can act as building blocks in constructive operations, namely, that parts can be combined in different ways instead of being driven deterministically towards unique aggregates.

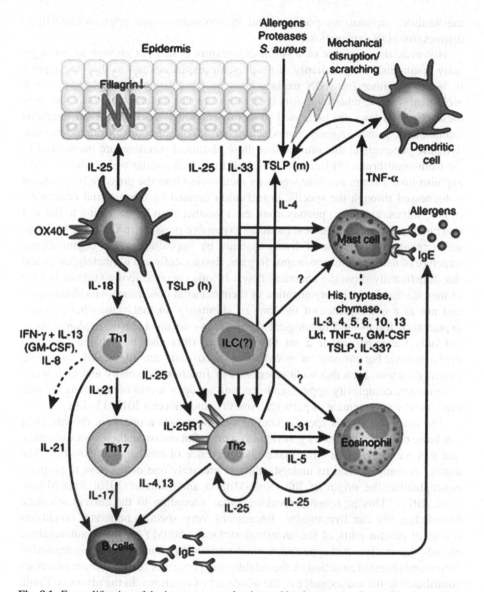

Fig. 8.1 Exemplification of the immune cascade triggered by the process of scratching in patients with dermatitis. The initial trigger, exemplified as a *mechanical disruption*, generates a reaction in different types of cells, which is partially caused by the allergens that bypass the epidermis during initial disruption. The mechanism is complex, involves several entities, and includes cases of loops, but it is mechanistic for it shows how the parts interact after an initial stimuli (scratch) triggers the reaction of the system. The final state is the inflammation of the epidermis. (From Cevikbas and Steinhoff 2012)

mechanistic explanations possible, and ontologically—and epistemologically—distinct from DN explanations.

However, the existence of a set of constraints is still not enough to ontologically distinguish the peculiarity of mechanistic explanations—as they are applied in biology—either from DN explanations, or from structural explanation. In a biological system, what biologists care about are not the chemical laws, but how these chemical laws are harnessed so as to generate a set of biological functions (metabolism, growth, agency, reproduction, etc.). And of course, from the thermodynamic perspective, this implies that these chemical reactions are maintained in far-from-equilibrium (FFE, hereafter) conditions. In a similar way as the engineer explains how a steam machine works by mentioning how the pressure of the steam is harnessed through the specific organization created by the material constraints (pipes, valves, cylinders, pistons, rods, etc.)—rather than by appealing to the fact that chemical laws say that heat produces expansive pressure in steam—a biologist will explain the behavior of living systems by appealing to the organizational structure of their internal constraints (organs, tissues, cells and ultimately, enzymes) that functionally drive the chemical flows. In other words, living systems behave as they do thanks to the organization of their material constraints—mechanisms—and not as a consequence of the laws of chemistry. In that vein, what provides explanatory force—and ontological distinctness—to mechanistic explanations is not only the existence of a set of constraints (this also happens in structural explanations), but *the way in which these constraints are specifically organized in living systems* such that wide repertoires of functional diversity and high levels of structural complexity appear, and generate an open world of functions through processes of arrangement of parts (cf. Arnellos and Moreno 2012, 15–18).

The existence of this specific kind of systems raises a question, though. How can basic chemical processes generate structures that constrain their own dynamics, and that we argue to be what makes the existence of mechanisms possible? The answer is complex, and its understanding is precisely one of the keys to properly conceptualize the origin of life (Ruiz-Mirazo and Moreno 2016; Ruiz-Mirazo et al. 2017). This is, roughly speaking, what according to the current scientific knowledge, we can hypothesize. Because of very specific boundary conditions (those of certain parts of the terrestrial surface 3.500MYA), many autocatalytic closed organizations had appeared in which certain compounds and/or aggregated of compounds exerted an action on the neighboring processes such that these processes contributed to the maintenance of the whole set of reactions. In the primitive Earth, driven by the energy of the Sun or by geo-thermal energy, many chemical reaction cycles would have appeared. For instance, a constant flow of energy and microporous surfaces would have favored the appearance of FFE chemical cycles leading to the formation of relatively complex organic compounds (Martin and Russell 2003). The idea of reaction cycle implies gathering together different reactions, i.e., embedding the processes of synthesis—and degradation—of new structures in a self-maintaining organization. The "core" of a chemical self-maintaining organization is the idea of *autocatalytic cycle*. Autocatalytic cycles are minimal forms of FFE chemical organizations. These systems are not only driven by external boundary conditions: a component—a catalyst—drives the network kinetically, keeping these systems in FFE conditions. A catalyst acts thus as a constraint: a material structure that selectively biases the microscopic lawful processes (in this case, a molecule which, because of its shape, selectively modifies the rates of certain reactions). The crucial point here is that an autocatalytic cycle generates a constraint, since the outcome of the action of the catalyst (the constraint) leads to its own re-production. The cycle therefore is not only maintained by the external set of boundary conditions, but also by the action of the constraint harnessing the underlying chemical interactions so that the process closes itself recursively. In more complex autocatalytic self-maintaining systems, it is possible to detect several constraints (catalysts, a membrane, etc.) in action. These diverse constraints should mutually enable their continuous regeneration, generating a cyclic causal regime called "closure of constraints" (cf. Mossio and Moreno 2010; Arnellos and Moreno 2012; Moreno and Mossio 2015). All this shows that the causal regime of biological systems is radically different from that of physico-chemical ones, in the sense that they create a set of material structures that constrain the underlying thermodynamic processes in a recursive way.

Importantly, we must highlight at this point that the initial appearance of constraints does not immediately lead to the origin of mechanisms, as the ontological conditions required for mechanisms are more demanding than the mere existence of a closure of constraints. As argued in Arnellos and Moreno (2012), the first protocells must have been able to exhibit minimal functional differentiation, realized by a system under a closure of constraints with the capacity of self-maintaining its organization. For this to happen, the protocell needs to start generating the basic components of its membrane, so as to keep it separate from its surrounding environment, and to guarantee that the processes that allow its self-maintenance to occur within its boundaries. However, this is still not enough to talk about mechanisms, for mechanisms demand further ontological requirements to be met. Concretely, they demand that some of the primitive set of constraints are organized to produce new functions based on combinatorial processes. That is to say, it is necessary that a very specific type of macromolecular components appear, so that they can play the role of constraints across different and potentially indefinite systems. Only once this happens, the kind of structural complexity that is achieved is such that it can trigger new structural changes and new combinatorial possibilities. And for this to happen, it is necessary that a modular organization emerges, i.e. *a type of organization in which different functional components may be separated and recombined allowing this wide arrange of combinational processes* (cf. Keller 2009). This is the basis for the generation of composite devices able to perform highly specific types of work, namely, *machines* (cf. Militello and Moreno 2018). Recall that, as it has been defined, a mechanism is a structure that performs a function in virtue of its entities, the activities they engage in, and their organization. The orchestrated functioning of the mechanism is responsible for one or more phenomena. Now, it is easy to see that many macromolecular aggregates in the cell fulfill this definition. For example, many proteins are constituted by secondary structures which are organized in rigid parts, whose relative movements can generate different effects, leading to the amplification of small displacements, or other mechanical effects described by a list of words borrowed from the description of machines: lever and spring, ratchet and clamp, etc. (Morange, personal communication). Of course, these material structures could be seen as machines performing a function only within the global organization of the cell.

In sum, though the appearance of systems whose global behavior can be explained through mechanisms requires a relatively high level of underlying complexity, the mechanistic description may ignore a lot of this underlying complexity to provide a successful form of explanation *at the relevant level*. This form of explanation is possible due to the existence of constraints that harness the physico-chemical laws, and to the existence of a global organization of constraints that performs a specific function and allows the regeneration of the set of constraints.

8.4 The Ontology of Structural Explanations: From Mechanisms to Networks and Back

Despite the importance of mechanistic explanations in biology, they cannot fully account for the biological phenomenology. As we said in Sect. 8.2, biological systems are constituted by an enormous amount of interacting entities. Everywhere and at every level of the biological organization, one can observe large sets of entities interacting in parallel (molecules in metabolisms, genes interacting in regulatory genetic networks, neurons connecting each other in neural networks, organisms exchanging chemicals with their neighbors in swarms, species interacting through food webs in ecosystems, etc.). These locally relatively simple interactions, however, often bring forth complex emergent behaviors. And in all these cases, what happens globally cannot be explained through functional decomposition, as we showed for the case of the microbiome, where the emergent stability behavior is not explained because of the causal relationships between the species that compose it, but as a result of the topological structure of the system. Indeed, the set of interacting species generates the global behavior, and thus the emergent property; but it is not possible to explain this behavior by attributing specific functions to each bacterium, or to each species, for many of them can be eliminated (perturbing the system) and the microbiome will keep exhibiting the same behavior.

We refer to these systems as "networks". By the concept of network is meant a system constituted by a set of units that interact locally in a simple way, where each unit is altered by interactions with its neighbors, and conversely, such that ultimately all units receive influences from all the other units of the system (Fig. 8.2). The defining characteristic of a network is the existence of recursion, which implies a closure of the paths of interaction on themselves. For example, a social network is a way of representing a social structure, assigning a graph: if two elements of the set of actors (such as individuals or social organizations) are related according to some criterion (professional relationship, friendship, kinship, etc.), then a line

Fig. 8.2 Graphical
representation of a random
network. The dots represent
the *nodes*, and the blue lines
that join them represent the
edges. Different networks
have different structure but
can be equally used to
represent different complex
systems (From Javaheripi
et al. 2019) (Color figure
online)

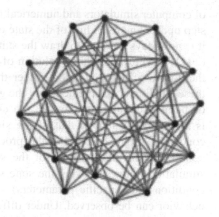

connecting the nodes representing these elements is constructed. Another example is metabolism, which can be viewed as a network. Metabolism could be represented by a system of units (nodes) interconnected (reactions) where nodes are metabolites and reactions are arcs with associated functions. A characteristic of the networks is the emergence of global patterns. These global regularities arise when the entire network is stabilized (a "dynamic attractor"), and in turn the emergence of these global phenomena may affect the behavior of the components of the network.

One of the main features of networks is their holism or systematicity, namely, that their global behavior cannot be explained through the detailed knowledge of the behavior of each unit (Jones 2014; Issad and Malaterre 2015; Green and Jones 2016; Deulofeu et al. 2019). Though authors like Bertalanffy have emphasized the importance of holistic processes in biological systems, the mechanistic atomistic strategy was predominant, because until recently no quantitative methods to deal with complex holistic systems were available. Fortunately, as we pointed out in Sect. 8.2, the study of the networks has been developed in the last decades thanks to the tools provided by computer simulation, as well as other mathematical developments. As a result, we have nowadays a new scientific domain called "graph theory" that studies the relation between network architectures and dynamics. This new scientific domain has allowed scientists to study complex holistic systems with strongly and recurrently interacting components showing that, despite their variety, they share certain generic properties (Moreno et al. 2011).

The scientific approach of complex holistic systems consists essentially in the quantitative study of the conditions under which certain sets of components, once they overcome a point or critical mass of interconnections, lead to the emergence of new global properties. This phenomenon happens in a wide variety of (physical, chemical, biological, neural, ecological, social, computer, etc.) systems, which can be said to be almost universal. But what is of particular interest is that, though the complexity of the dynamics leading to the global behavior of the network is not analytically tractable, there are ways to predict that behavior that rely solely on the level of interconnectedness among the parts of the system. In fact, thanks to the use of computer simulators and numerical methods consisting in a fine-grained step-by-step update and recording of the state of all the interrelated variables of the system, it is nowadays possible to draw the state space of these networks. As Moreno et al. (2011, 318) explain, "the evolution of the system is "synthetically" reproduced in the course of the simulation, rather than deduced for a given time value. Under these conditions, the prediction of the global property requires an enormous amount of parallel computation, where the complexity of the computational simulation is almost equivalent to that of the simulated system. Since sensitivity to initial conditions might be critical, the process is repeated for a wide range of initial conditions and the behavior of the system recorded. A full record of different simulations allows drawing the state space of the system (given certain boundary conditions and specific parameters) so that regular and reproducible patterns of behavior can be observed. Under different network configuration parameters, the rules governing components or boundary conditions can be systematically studied through simulation, and a deeper understanding of the structure-function mapping is gained without a strict decomposition or a one-to-one localization."

As we said in Sect. 8.2, the explanatory strategy that relies on the network structure of the systems is called "structural explanation" (after Huneman 2018a). Epistemologically, and in contrast with mechanistic explanations, structural explanations do not gain their explanatory force from unveiling the causal (mechanistic) structure of a given phenomenon (or rather, a set of similar phenomena). Rather, structural explanations rely solely on the mathematical organization of the system and presume that the entities and activities will behave in such a way that the phenomenon to explain will obtain, provided that their specific organization is not altered. Ontologically, systems that realize a network-structure *constrain* the behavior of the entities that compose them so that the emergent properties that the system realizes are maintained because the recursive/cyclic action of these constraints, as we explained in Sect. 8.3 (see also Arnellos and Moreno 2012, 15–18). The nature of the entities in the system, the type of activities they perform, or the causal paths that they follow are largely irrelevant and, importantly, can be very different across empirical systems that share the same structural explanation, while the final outcome—an emergent and systemic property—does not get altered provided that the organization (or "architecture") of the system is the same across systems. In that vein, a structural explanation results from the identification of an empirical system with a mathematical structure, and from the ascription of the properties of the mathematical structure to every empirical system that can be ascribed the same mathematical structure (Huneman 2010, 2018a, b; Jones 2014; Deulofeu et al. 2019).

Structural explanatory strategies, in their specific application to network models, can also be considered in a different way, namely, as a tool for predicting the behavior of complex dynamic systems. In this sense, and again in sharp contrast with mechanistic explanations, *network models provide a prediction* of the global behavior of a complex holistic system, rather than an understanding of the relations and functions of their parts—entities, activities. Of course, this second sense is not completely different from the first one, since the behavior of the network is a consequence of its mathematical properties. The difference lies in the fact that whereas the former sense looks for the establishment of similitudes between different empirical systems that have the same mathematical structure (e.g. computers, socials networks, genetic networks, and metabolic networks), the second is more focused on the study of the specific dynamical properties of an empirical complex system (e.g. the microbiome, or the immune system).

This immediately raises a question. What is the ontological reason that makes networks suitable systems for structural explanations? In a sense, the functioning of complex holistic networks can be more easily related to a traditional law-based physical system than to a mechanistic organization. Of course, in a network we have a huge number of observables and, rather than a universal law, we can have different local rules, depending on the type of network we are dealing with—random, small-world, scale-free, etc. The behavior of a complex network cannot be described in a simple set of differential equations, but, as Fox Keller has pointed out, "in a messy complex of algorithms, vast systems of differential equations, statistical analyses, and simulations. Such models can only be successfully formulated in the most intimate back and forth relation with experiment", (Keller 2005a, 7–8). Despite these mathematical difficulties, a peculiarity of network analysis is that the behavior of the whole system can be calculated, and therefore predicted. This prediction is possible because the level of interconnectedness of the elements within the network, and the way in which they are globally related, generates a structure whose behavior becomes expectable irrespectively of the nature of the entities that interact, or the specific causal paths that they follow. In other words, structural explanations that rely on the network-structure of the system work in virtue of the role of the system in constraining the behavior of the entities that constitute them in a recursive way.[6]

Concerning this last point, William Bechtel and some of his collaborators have recently argued that network explanations are not, in the end, different from mechanistic explanations, provided that the concept of mechanism, and the heuristics dictated by the mechanistic explanatory strategy are understood correctly (Bechtel and Abrahamsen 2005; Bechtel 2015a, b). Bechtel argues that, if we recognize that the boundaries of the mechanism are always imposed by the researcher, we can see that network explanations are not but ways of expanding the boundaries of the mechanism in the search of additional mechanistic components at any level. To quote: "In such cases, [scientists] can relax these boundaries and search for additional entities outside the mechanism as initially construed with which its parts interact or consider longer timescales during which the operations have effects. There is no guarantee that this strategy will always be successful, but when it is, the pursuit of mechanistic explanation is still a useful research strategy. It is important, though, to keep in mind that *it is a heuristic strategy of scientists who delineate mechanism boundaries*" (Bechtel 2015b, 93, emphasis ours).

---

[6]Notice that this requirement is less strict than the ontological requirements to speak about mechanisms, as we said in Sect. 8.3.

We agree with Bechtel that the mechanistic heuristic can be applied to the study of systems that operate over longer timescales than those of simple, standard mechanisms. Furthermore, we can concede that part of this is what happens in some cases of network analysis (e.g. prediction of the behavior of specific empirical complex systems) insofar as the model of the mechanism seems to be always included in at least some of these explanations, as argued in Deulofeu et al. (2019). However, there is a substantial difference that Bechtel does not mention, and that generates a clear difference between the appeal to mechanisms and the appeal to networks in scientific explanation. While in the case of the former, the existence of a specific causal story that brings the mechanisms from a set of initial conditions to a termination condition is necessary, or there is not mechanistic explanation, the same is not true in the case of structural explanations that appeal to the network structure of the system. In the case of the latter, the causal details are irrelevant to provide a successful explanation, and the reason why this is so, we argue, is ontological: in a network system, the set of constraints that is created by the global architecture—conceived in terms of interconnectedness—is such that the global property that emerges can be realized by a set of different causal configurations. This means that even when a network contains mechanisms, their role could be ignored, since the same type of properties could correspond to a variety of underlying mechanisms, as these are individuated according to the specific causal stories that define them.

In that vein, however, network-based analyses do not allow specific interventions in the systems that they explain, for they remain neutral with respect to the causes that bring the phenomena about. It is for this reason that the discovery of the specific type of mechanism becomes necessary to understand how and why a given biological system behaves, as well as to discover specific interventions in the system. But also, notice that despite this, there are ontological reasons to consider these systems different from mechanistic systems. As Fox Keller has argued, "Biological systems are, as we know, extraordinarily complex, but again because of evolution, they are complex in somewhat different ways than systems in physics are understood to be complex: for one, they are always and inevitably hierarchical. Accordingly, familiar notions of emergence, rooted in the non-linear dynamics of uniform systems (gases, fluids, or lattices), are not adequate to the task (. . .) The central point is that the inhomogeneities and ordered particularities of biological systems are essential to their functioning and hence cannot be ignored; indeed, to ignore them is to risk exactly the kind of biological irrelevance that has historically been the fate of so many mathematical models in biology" (Keller 2005a, 7). And, referring explicitly to the explanatory scope of structural explanations, she argues that "first, power law distributions are neither new nor rare; second, fitting available data to such distributions is suspiciously easy; third, even when the fit is robust, it adds little if anything to our knowledge either of the actual architecture of the network, or of the processes giving rise to a given architecture (many different architectures can give rise to the same power laws, and many different processes can give rise to the same architecture" (Keller 2005b, 1066).

In the same line, Huneman admits that structural explanations are incomplete and need to be complemented by mechanistic explanations—and the other way around. To quote: "So a first kind of relation between mechanistic and topological explanations is that they can be two stages of a complete explanation of the same phenomenon, related in a diachronic way: why does phenomenon X exists? Because of some topological properties Ti of the system X; and why does S have Ti? Because of some mechanism proper to S, its surroundings and parts, etc."[7] (Huneman 2010, 226)

Accordingly, there are good reasons to look both for mechanisms and for network structures if we try to fully understand biological phenomena.

8.5 Biological Systems Do Require Complex Explanations: How Mechanistic and Structural Explanations Relate

In fact, neither mechanistic nor structural explanations, taken separately, are sufficient to fully understand why biological systems behave the way they do. Let us take the example of robustness. Robustness refers to the ability of a system to keep its properties—e.g. its organization—despite the perturbations it experiences. Very often, network-like models are provided to explain the degree of robustness of a great variety of biological systems (ecosystems, metabolic networks, developmental processes, the microbiome, etc.). On the other hand, at the same tyme, more local analyses tend to focus on the set of specific mechanisms that causally try to explain why a given type of organization is more robust against perturbations than another one.

But how are these two approaches related in biological systems? In other words, how both of them—being radically different—could notwithstanding provide a valid explanation of some of the properties of these systems? Ultimately, network like models explain robustness by redundancy (cf. Kitano 2004) and/or by "distributed robustness" (cf. Wagner 2005), namely, because the network is such that it contains different and alternative pathways which buffer the outcome despite possible change in one of the nodes. Yet, behind this level of description, a regulatory mechanism could be in place. For example, in the development of full-fledged multicellular organisms, a gene regulatory subsystem exerts a fine-tuned control on how and when early-undifferentiated cells become differentiated and spatially allocated. Notice that there are many examples of multicellular systems genetically homogeneous presenting a certain degree of cellular differentiation without the existence of any (developmental) regulatory subsystem. The problem, however, is that functional coordination is achieved only if the functional diversity is poor, and these multi-

---

[7]Notice that Huneman uses "topological" to refer to a specific subclass of structural explanations, as he makes clear in Huneman (2018a).

cellular organizations cannot achieve significant degrees of functional integration.[8] Now, although a regulatory mechanism is (as any other mechanism) obviously based on localized functional parts, it could happen that certain parts of the mechanism constitute themselves a network. This is the case of the regulatory genetic networks. To understand why they constitute a mechanism, we have to understand that this epigenetic molecular network constrains functionally the intercellular relations. And it can do this task because it is organized in a special way, namely, operating in a dynamically decoupled way from the level of the processes they constrain according to a higher-level norm (the epigenetic program) (cf. Bich et al. 2016). But considering the epigenetic domain, these regulatory genes interact with each other recurrently, bringing forth certain topological (structural) properties which, substituting localized action, become the relevant level of control. Thus, in this example, we see that it is the articulation between a functionally distinguishable organization of constraints—the regulatory mechanism—and a distributed recurrent network of (part of) these constraints that, all together, constitute the explanation of how functionally diversified development happens.

This shows that the higher degree of complexity achieved by biological systems requires an organization structured in different (sub)systems interconnected at different levels and time scales. In other words, it requires the creation, within the system, of subsystems that can modify the parameters of other subsystem parts. Since these subsystems work at different rates and with different operational rules, the system has an increased potential to explore new alternative forms of global self-maintenance that are not accessible to 'flat' systems without any hierarchy or modularity in their organization. In this way, the higher-level subsystem creates a set of functional constraints on the lower-level dynamics. At the same time, the controlled level plays a fundamental role in the constitution and maintenance of the controller level (and hence, of the whole system).

Therefore, biological systems convey specific forms of complexity that, through holistic-emergent processes (which are continuously taking place), produce both network holistic patterns and new, more complex structures which, in turn, are bound to become selective functional constraints acting on the dynamic processes that underlie those holistic processes. The reason why those functional constraints can be described as mechanisms is that they act as distinguishable parts (or collections of parts) related to particular tasks (e.g., catalytic regulation) performed in the system. So, both aspects are, thus, complementary: the holism of the global network of processes and the local control devices/actions that are required for the system to increase its complexity. Moreover, the newly created and functionally diverse constraints may give rise (once a certain degree of variety is reached) to new self-organizing holistic processes, which, in turn, may be functionally re-organized (cf. Moreno et al. 2011). In this way, an increase in organizational complexity can take the paradoxical form of an apparent "simplification" of the underlying

---

[8]For a discussion on this point, see Arnellos et al. (2014). For a general discussion on regulation and complex functional integration, see Christensen (2007) and Bich et al. (2016).

complicatedness, giving rise to levels of organization in which a mechanistic decompositional strategy might be locally applicable. The idea, taken originally from Pattee (1973), would be that new hierarchical levels are created through a functional loss of details of the previous ones, thus generating systems where both mechanisms *and* networks are simultaneously realized and co-existent.

8.6 Conclusions

The organization of biological systems combines holistic processes described in terms of networks and others with distinguishable parts that can be described as mechanisms. What is characteristic of the biological complexity is precisely the entanglement of processes that require locally mechanistic explanations, with others, more global, massively parallel and holistic, whose behavior tends to generate "emerging patterns" and phenomena of "self-organization". This combination can occur at various levels, and can generate new forms of holism, because interactions between functionally distinct parts in a lower level can generate global emerging phenomena in a higher level, and conversely. Therefore, the study of biological systems requires a deep reconsideration of the approaches and methodologies developed so far, aiming at an integration of the mechanistic methodology, based on the analysis of functional parts and the reconstruction of behaviors, with the new network-modeling, based on the discovery of forms of regularity in massively parallel non-linear systems.

It is time to come back to Nagel's challenge: if, as modern science argues, the grounding of complex level phenomena are but physical systems, science has to explain the relevance of specific forms of explanations for higher level phenomena, namely, why these specific forms of explanations work, and how they are related with the DN type of explanations that are still at work successfully in the "fundamental" sciences. After all, we could not respond to the question of why purely physical elements, when combined in a certain way, can give raise to biological (or mental and social) properties in a "non-magic" way (as Nagel says) only by showing that we have specific ways of explanation for biological, mental or social phenomena that work successfully. It is necessary also to explain how and why these explanations are connected with each other and, above all, to explain how and why they could be connected with the basic explanations of the physical sciences. In this chapter, we have started this task by exploring: first, how the ontology that grounds the validity of mechanistic vs. structural explanations differs, and how this results from the evolution of living systems (Sects. 8.3 and 8.4); second, by arguing how and why these two modes of explanations relate to each other, and explaining why they are sometimes complementary (Sect. 8.5); finally, by connecting all this to the way in which living systems harness the laws of nature to create their own constraints, thus generating their own regime of causation.

Acknowledgments Alvaro Moreno acknowledges the financial support from the University of the Basque Country (PES 18/92), from the Basque Government (IT 1228-19), from the Spanish Ministry of Economía y Competitividad (MINECO, FFI2014-52173-P Grant) and from that of Ciencia, Innovación y Universidades (PID2019-104576GB-I00). Javier Suárez formally acknowledges the Deutsche Forschungsgemeinschaft (DFG, German Research Foundation) in the project "Complex Biological Dispositions: A Case Study in the Metaphysics of Biological Practice" (288923097) as part of the research group "Inductive Metaphysics" (FOR 2495), and the Spanish Ministry of Economy and Competitiveness (FFI2016-76799-P) for the funding that made this research possible.

References

Alleva, K., Díez, J., & Federico, L. (2017). Models, theory structure and mechanisms in biochemistry: The case of allosterism. *Studies in History and Philosophy of Biology and Biomedical Sciences, 63*, 1–14.

Arnellos, A., & Moreno, A. (2012). How functional differentiation originated in prebiotic evolution. *Ludus Vitalis, 37*, 1–23.

Arnellos, A., Moreno, A., & Ruiz-Mirazo, K. (2014). Organizational requirements for multicellular autonomy: Insights from a comparative case study. *Biology and Philosophy, 29*(6), 851–884.

Alexander, S. (1920). *Space, time and deity*. London: Macmillan.

Bechtel, W. (2015a). Can mechanistic explanation be reconciled with scale-free constitution and dynamics? *Studies in History and Philosophy of Biology and Biomedical Sciences, 53*, 84–93.

Bechtel, W. (2015b). Generalizing mechanistic explanations using graph-theoretic representations. In P. A. Braillard & C. Malaterre (Eds.), *Explanation in biology: An enquiry into the diversity of explanatory patterns in the life sciences* (pp. 199–225). Dordrecht: Springer.

Bechtel, W., & Abrahamsen, A. (2005). Explanation: A mechanist alternative. *Studies in History and Philosophy of Science Part C: Studies in History and Philosophy of Biological and Biomedical Sciences, 36*(2), 421–441.

Bechtel, W., & Richardson, R. C. (1993). *Discovering complexity: Decomposition and localization as scientific research strategies*. Cambridge, MA: The MIT Press.

Bich, L., Mossio, M., Ruiz-Mirazo, K., & Moreno, A. (2016). Biological regulation: Controlling the system from within. *Biology and Philosophy, 31*(2), 237–265.

Brigandt, J., Green, S., & O'Malley, M. A. (2017). Systems biology and mechanistic explanation: Chapter 27. In S. Glennan & P. Illari (Eds.), *The Routledge handbook of mechanisms and mechanical philosophy* (pp. 362–374). London: Routledge.

Broad, C. D. (1925). *The mind and its place in nature*. London: Routledge and Kegan Paul Ltd..

Cevikbas, F., & Stenhoff, M. (2012). IL-33: A novel danger signal system in atopic dermatitis. *Journal of Investigative Dermatology, 132*(5), 1326–1329.

Christensen, W. (2007). The evolutionary origins of volition. In D. Ross, D. Spurrett, H. Kincaid, & L. Stephens (Eds.), *Distributed cognition and the will: Individual volition and social context* (pp. 255–287). Cambridge, MA: MIT Press.

Craver, C. F. (2007). *Explaining the brain*. New York: Clarendon Press.

Craver, C. F. (2014). The ontic account of scientific explanation. In M. I. Kaiser, O. R. Scholz, D. Plenge, & A. Hüttemann (Eds.), *Explanation in the special sciences: The case of biology and history* (pp. 27–52). Springer: Dordrecht.

Craver, C. F., & Darden, L. (2013). *In search for mechanisms: Discovery across the life sciences*. Chicago: University of Chicago Press.

Deulofeu, R., & Suárez, J. (2018). When mechanisms are not enough: The origin of eukaryotes and scientific explanation (chapter 6). In A. Christian, D. Hommen, N. Retzlaff, & G. Schurz (Eds.), *Philosophy of science* (Vol. 9, pp. 95–115). Cham: Springer.

Deulofeu, R., Suárez, J., & Pérez-Cervera, A. (2019). Explaining the behaviour of random ecological networks: The stability of the microbiome as a case of integrative pluralism. *Synthese*, pp. 1–23 https://doi.org/10.1007/s11229-019-02187-9. Accessed 7 Aug 2019.

Díez, J. A. (2014). Scientific W-explanation as ampliative, specialised embedding: A neo-Hempelian account. *Erkenntnis, 79*, 1413–1443.

Díez, J. A., Khalifa, K., & Leuridan, B. (2013). General theories of explanation: Buyers beware. *Synthese, 190*, 379–396.

Dupré, J. (2013). Living causes. *Aristotelian Society Supplementary Volume, 87*(1), 19–37.

Dupré, J. (2017). The metaphysics of evolution. *Interface Focus, 7*(5). https://doi.org/10.1098/rsfs.2016.0148. Accessed 7 Aug 2019.

Giere, R. N. (1999). *Science without laws*. Chicago: University of Chicago Press.

Glennan, S. (1996). Mechanisms and the nature of causation. *Erkenntnis, 44*(1), 49–71.

Glennan, S. (2002). Rethinking mechanistic explanation. *Philosophy of Science, 69*(S3), S342–S353.

Green, S. (2017). Philosophy of systems and synthetic biology. In: E. N. Zalta (Ed.). *The Stanford encyclopedia of philosophy*. https://plato.stanford.edu/entries/systems-synthetic-biology/. Accessed 7 Aug 2019.

Green, S., & Jones, N. (2016). Constraint- based reasoning for search and explanation: Strategies for understanding variation and patterns in biology. *Dialectica, 70*(3), 343–374.

Hempel, C. (1965). *Aspects of scientific explanation and other essays in the philosophy of science*. New York: Free Press.

Humphreys, P. (1997). How properties emerge. *Philosophy of Science, 64*, 1–17.

Huneman, P. (2010). Topological explanations and robustness in biological sciences. *Synthese, 177*, 213–245.

Huneman, P. (2018a). Outlines of a theory of structural explanation. *Philosophical Studies, 175*(3), 665–702.

Huneman, P. (2018b). Diversifying the picture of explanations in biological sciences: Ways of combining topology with mechanisms. *Synthese, 195*, 115–146.

Huneman, P. (2018c). Realizability and the varieties of explanation. *Studies in History and Philosophy of Science*. https://doi.org/10.1016/j.shpsa.2018.01.004.

Issad, T., & Malaterre, C. (2015). Are dynamic mechanistic explanations still mechanistic? In P. A. Braillard & C. Malaterre (Eds.), *Explanation in biology: An enquiry into the diversity of explanatory patterns in the life sciences* (pp. 265–292). Dordrecht: Springer.

Javaheripi, M., Rouhani, B. D., & Koushanfar, F. (2019). SWNet: Small-world neural networks and rapid convergence. arXiv:1904.04862.

Jones, N. (2014). Bowtie structures, pathway diagrams, and topological explanations. *Erkenntnis, 79*(5), 1135–1155.

Keller, E. F. (2005a). The century beyond the gene. *Journal of Biosciences, 30*, 3–10.

Keller, E. F. (2005b). Revisiting "scale-free" networks. *BioEssays, 27*, 1060–1068.

Keller, E. F. (2009). Self-organization, self-assembly, and the inherent activity of matter. In S. H. Otto (Ed.), *The Hans Rausing lecture 2009*. Uppsala: Uppsala University, Disciplinary Domain of Humanities and Social Sciences, Faculty of Arts, Department of History of Science and Ideas.

Kim, J. (1989). The myth of non-reductive materialism. *Proceedings and Addresses of the American Philosophical Association, 63*(3), 31–47.

Kim, J. (1999). Making sense of emergence. *Philosophical Studies, 95*, 3–36.

Kim, J. (2006). Emergence: Core ideas and issues. *Synthese, 151*, 547–559.

Kitano, H. (2004). Biological robustness. *Nature Review Genetics, 5*, 826–837.

Kitcher, P. (1989). Explanatory unification and the causal structure of the world. In P. Kitcher & W. Salmon (Eds.), *Scientific explanation* (pp. 410–505). Minneapolis: University of Minnesota Press.

Krickel, B. (2018). *The mechanical world: The metaphysical commitments of the new mechanistic approach*. Dordrecht: Springer.

Lloyd Morgan, C. (1923). *Emergent evolution*. London: Williams and Norgate.

Machamer, P., Darden, L., & Craver, C. (2000). Thinking about mechanisms. *Philosophy of Science, 67*, 1–25.

Martin, W., & Russell, M. J. (2003). On the origins of cells: A hypothesis for the evolutionary transitions from abiotic geochemistry to chemoautotrophic prokaryotes, and from prokaryotes to nucleated cells. *Philosophical Transactions of the Royal Society: Biological Sciences, 358*, 59–85.

McLaughlin, B. (1992). The rise and fall of British emergentism. In A. Beckerman, H. Flohr, & J. Kim (Eds.), *Emergence or reduction? Essays on the prospects of non-reductive physicalism* (pp. 49–93). Berlin: De Gruyter.

McLaughlin, B. (1997). Emergence and supervenience. *Intellectica, 2*(25), 25–43.

McManus, F. (2012). Development and mechanistic explanation. *Studies in History and Philosophy of Biology and Biomedical Sciences, 43*, 532–541.

Mill, J. S. (1843). *A system of logic*. London: Parker.

Militello, G., & Moreno, A. (2018). Structural and organizational conditions for being a machine. *Biology and Philosophy, 33*, 35.

Mitchell, S. D. (1997). Pragmatic Laws. *Philosophy of Science, 64*, S468–S479.

Mitchell, S. D. (2003). *Biological complexity and integrative pluralism*. Cambridge: Cambridge University Press.

Mitchell, S. D. (2012). Emergence: Logical, functional and dynamical. *Synthese, 185*(2), 171–186.

Moreno, A., Ruiz-Mirazo, K., & Barandiaran, X. E. (2011). The impact of the paradigm of complexity on the foundational frameworks of biology and cognitive science. In C. A. Hooker, D. V. Gabbay, P. Thagard, & J. Woods (Eds.), *Handbook of the philosophy of science* (Philosophy of complex systems) (pp. 311–333). Oxford: Elsevier.

Moreno, A., & Mossio, M. (2015). *Biological autonomy: A philosophical and theoretical enquiry*. Dordrecht: Springer.

Mossio, M., & Moreno, A. (2010). Organizational closure in biological organisms. *History and Philosophy of the Life Sciences, 32*(2–3), 269–288.

Nagel, T. (2012). *Mind and cosmos. Why the materialist neo-Darwinian conception of nature is almost certainly false*. Oxford: Oxford University Press.

Nicholson, D. J. (2012). The concept of mechanism in biology. *Studies in History and Philosophy of Biology and Biomedical Sciences, 43*, 152–163.

Nicholson, D. J. (2013). Organism ≠ Machines. *Studies in History and Philosophy of Biology and Biomedical Sciences, 44*, 669–678.

Nicholson, D. J. (2018). Reconceptualizing the organism: From complex machine to flowing stream. In D. J. Nicholson & J. Dupré (Eds.), *Everything flows: Towards a processual philosophy of biology* (pp. 139–166). Oxford: Oxford University Press.

Pattee, H. H. (1972). Laws and constraints, symbols and languages. In C. Waddington (Ed.), *Towards a theoretical biology* (Vol. v. 4, pp. 248–258). Edinburgh: Edinburgh University Press.

Pattee, H. H. (1973). The physical basis and origin of hierarchical control. In H. Pattee (Ed.), *Hierarchy theory. The challenge of complex systems* (pp. 73–108). New York: George Braziller.

Polanyi, M. (1968). Life's irreducible structure. *Science, 160*, 1308–1312.

Rice, C. (2015). Moving beyond causes: Optimality models and scientific explanation. *Noûs, 49*(3), 589–615.

Ruiz-Mirazo, K., & Moreno, A. (2016). Reflections on the origin of life: More than an evolutionary problem. *Mètode Science Studies Journal, 6*, 151–159.

Ruiz-Mirazo, K, Briones, C., & de la Escosura, A. (2017). Chemical roots of biological evolution: The origins of life as a process of development of autonomous functional systems. *Open Biology*. https://doi.org/10.1098/rsob.170050. Accessed 7 Aug 2019.

Salmon, W. (1984). *Scientific explanation and the causal structure of the world*. Princeton: Princeton University Press.

Suárez, J., & Deulofeu, R. (2019). Equilibrium explanation as structural non-mechanistic explanations: The case of long-term bacterial persistence in human hosts. *Teorema, 38*(3), 95–120.

Suárez, J., & Triviño, V. (2019). A metaphysical approach to holobiont individuality: Holobionts as emergent individuals. *Quaderns de Filosofia, 6*(1), 59–76.

Umerez, J., & Mossio, M. (2013). Constraint. In W. Dubitzky, O. Wolkenhauer, K. Cho, & H. Yokota (Eds.), *Encyclopedia of systems biology* (pp. 490–493). Dordrecht: Springer.

Wagner, A. (2005). *Robustness and evolvability in living systems*. Princeton: Princeton University Press.

Wilson, J. (2013). Nonlinearity and metaphysical emergence. In S. Mumford & M. Tugby (Eds.), *Metaphysics and science* (pp. 201–235). Oxford: Oxford University Press.

Wilson, J. (2016). Metaphysical emergence: Weak and strong. In T. Bigaj & C. Wüthrich (Eds.), *Metaphysics in contemporary physics* (pp. 345–402). Boston: Brill Rodopi.

Witherington, D. (2011). Taking emergence seriously: The centrality of circular causality for dynamic systems approaches to development. *Human Development, 54*, 66–92.

Woodward, J. (2000). Explanation and invariance in the special sciences. *British Journal for the Philosophy of Science, 51*, 197–254.

Woodward, J. (2003). *Making things happen: A theory of causal explanation*. New York: Oxford University Press.

Woodward, J. (2017). Scientific explanation. In: E. N. Zalta (Ed.). *The Stanford Encyclopedia of philosophy*. https://plato.stanford.edu/archives/fall2017/entries/scientific-explanation/. Accessed 7 Aug 2019.

Chapter 9
Scientific Prediction and Prescription in Plant Genetic Improvement as Applied Science of Design: The Natural and the Artificial

Pedro Martínez-Gómez

Abstract Natural sciences and social sciences are typically included within the empirical sciences. However, we can also consider a different type of empirical science: the sciences of the artificial. In this context, Plant Genetic Improvement (PGI) has a double scientific nature: it is both an empirical science of nature (analysing the natural genetic variation in the plant kingdom) and a science of the artificial (*human-made*) as an applied science of design for the development of new releases or plant varieties. This paper provides an analysis of scientific prediction as an essential feature of PGI as an applied science of design, giving support to the existence of a diversity of methods in PGI.

This methodological variety is open to a number of consequences, such as differences in the research according to the three levels of knowledge associated with PGI: molecular biology (micro level); the genetic constitution of the individuals (meso level); and the phenotype and overall appearance of the newly designed individuals (released) (macro level). These levels of reality affect both the type of prediction and the methodology applied associated with prescription in PGI as an as applied science of design.

Regarding prediction as the main objective in PGI, it is very important to have suitable knowledge of the possible future to make a new variety through human design. In addition, in developing a scientific prediction applied to PGI, it is necessary to consider different internal (the genetic nature of the starting plant material, the available methodologies, etc.) and external (social acceptance, environmental factors, biotic and abiotic stresses, etc.) variables. The degree of knowledge of these variables determines the quality of the prediction in the design of new plant varieties.

P. Martínez-Gómez (✉)
Departamento de Mejora Genética, CEBAS-CSIC, Murcia, Spain

Instituto de Estudios de la Ciencia y la Tecnología, Universidad de Salamanca, Salamanca, Spain
e-mail: pmartinez@cebas.csic.es

© Springer Nature Switzerland AG 2020   167
W. J. Gonzalez (ed.), *Methodological Prospects for Scientific Research*, Synthese
Library 430, https://doi.org/10.1007/978-3-030-52500-2_9

Keywords Plant genetics improvement · Science of the artificial · Design science · Prediction · Prescription · Internal variables · External variables

9.1 Plant Genetic Improvement as an Empirical Applied Science and Science of Design

The kind of knowledge involved in any given experiment is central in the distinction between basic sciences and applied sciences. Basic sciences are focused on expanding scientific knowledge regarding reality, whereas applied sciences use a type of research oriented toward solving specific problems within a given domain (cf. Niiniluoto 1993). Both types of science follow different philosophical and methodological frameworks. In the first case, the task is to explain and predict reality, while in the second case, research is performed by predicting and prescribing. The applied sciences have a different modus operandi. They are oriented toward solving specific problems within a concrete sphere. These sciences are configured in a way that includes aims, processes and results.[1]

Natural sciences and social sciences are typically included within the empirical sciences. However, we can also consider a different type of empirical science: the sciences of the artificial (cf. Simon 1996). These sciences are in the realm of the *human-made* (cf. Gonzalez 2008). This means that there are objectives and specific processes for solving a particular problem. The *design* from human creativity guides the process for reaching objectives or goals. Within this group of disciplines, there is a specific type of science called *design science*. Sciences that fall under this category are applied sciences associated with aims, processes and results and that strive to obtain certain results from a specific design. Important examples of this branch of the sciences of the artificial, in general, and design sciences, in particular, can be found in research in the fields of economy, pharmacology and computer science.[2]

9.1.1 Configuration of Plant Genetic Improvement as an Empirical Science and an Applied Science of Design

Plant Genetic Improvement (onwards PGI) is an applied science of design. It requires empirical support based on observation and experimentation (cf. Gonzalez 2010a). This science also includes a component of nature. Research in PGI can thus be considered an empirical applied science. At the same time, PGI is a science of

---

[1] A full-fledged study of this topic can be found in Gonzalez (2007). Concerning the questions related to explanation in Science see Gonzalez (2002).

[2] The dynamic aspects of the sciences of the artificial, in general—and of the design sciences, in particular—in different fields are presented in Gonzalez (2013a, 2017a, 2018a) and Gonzalez and Arrojo (2019).

the artificial due to its aims, processes and results, which emerge from a design. PGI is thus the basis for biotechnology (cf. Jasanoff 2006) applied to the development of new *human-made* plant varieties. While this science is increasingly linked to enhancing human possibilities, which is one feature of the design sciences, it is also closely related to the development of new patents, and thus provides scientific knowledge that is the basis for the agricultural industry.

PGI is firstly an applied science, insofar as it is focussed on solving specific problems, such as productivity, resistance to biotic and abiotic stresses, fruit quality, etc. Secondly, PGI is a design science, as it comes up with models to meet aims that expand human possibilities. This requires processes that lead to results. In fact, the design sciences involve applied knowledge, because the models proposed have a practical dimension: they seek to solve specific problems (in the short, medium or long term) (cf. Gonzalez 2013a, b).

In addition, PGI has a link with technology since it can lead to innovation, an achievement produced by the creative transformation of reality.[3] This can lead, for example, to the creation of a new plant variety, which is a desired transformation. This is clearly different from a discovery. An innovation or *invention* is an outcome of materials and procedures that do not exist in nature: it requires the intervention of a human agent and the following of a plan (e.g., with new varieties of fruit, etc.). Thus, while a discovery is not patentable, an invention is.

PGI is thus a scientific discipline that contributes to the knowledge necessary to use a certain technology, which in turn applies that knowledge to produce new plant varieties and plant patents. This technological task is carried out via a timely action that changes reality. This concept of *technology* differs from the notion of *science*, because science does not seek to modify reality, but to enhance our knowledge of reality. Scientific activity and technological endeavours are two different things, but they have common connections. In fact, scientific knowledge and technology interact.[4] Therefore, through PGI as a dual science—a natural science and a design science—we can contribute to the basis for a certain *technology*. In this regard, knowledge as an applied science of design makes it possible to creatively transform reality and thus leads to producing a particular new product or artefact (cf. Gonzalez 2013a, b).

The emphasis on PGI as a human activity rather than mere content shows the advantages of pragmatism and pluralism as methodological prospects in scientific research. They can overcome the problems of monism, reductionism and methodological universalism as well as the proposals of methodological imperialism. This paper thus gives support to the existence of a diversity of methods in PGI as an applied science of design.

---

[3]A specific characteristic of technology is a creative transformation of the real. See Gonzalez (2005).

[4]See several examples of interaction between scientific knowledge and technology in Gonzalez (2005, 2013b, 2015a).

9.1.2   The Ontological and Epistemological Complexity of Plant Genetic Improvement as an Applied Science of Design: Structural and Dynamic Dimensions

The scientific status of PGI has two sides—it is both a natural science and a design science, and can thus be considered a complex science. PGI involves both dealing with an aspect of reality that is to be researched—the ontological component—and the approach to knowing that reality—the epistemological factor—. If the focus is on designs, it seems clear that they arise from a combination of different elements. There are two initial aspects of a design: (i) an interaction between parts within a whole, and (ii) in principle, a hierarchical relationship, when a system with internal articulation is configured.

New designs (in our case new varieties) are also oriented towards progressively more sophisticated aims. In fact, it is becoming increasingly difficult to compute all the available information for these new, more sophisticated designs, and the difficulties are even greater when the design is more ambitious. It is also necessary to take into account the functional complexity of the designs (cf. Gonzalez 2012a, 2013a).

The products generated by man using artificial criteria can be characterised as *bio-artefacts* (cf. Cuevas-Badallo 2008). We must also consider the existence of a dynamic structural complexity to be addressed by PGI as a science of the artificial. This structure can be considered by analogy with Herbert Simon's descriptions of the science of the artificial at the structural and dynamic levels (cf. Simon 1996).

From a structural point of view, within PGI, we would have an organised system that can be broken down into different subsystems. The structure has internal mechanisms with goal-oriented designs, which require certain processes to be developed and produce results that must be evaluated. It is necessary to select these objectives, processes and outcomes. Furthermore, in order to produce the design, besides scientific knowledge, we must also look to knowledge outside of science (e.g., practices that have produced results by trial and error, without a scientific basis) in addition to social or economic data (cf. Cuevas-Badallo 2005).

From a dynamic point of view, there is also a historical aspect in producing a new design or variety, because solving specific problems occurs within an ever-evolving environment historically speaking (cf. Gonzalez 2012b). The field of PGI as an applied science of design also involves complex dynamics. This discipline is being developed as an open process with many possibilities. This leads to an internal component in the dynamics—the aims, processes and results—and an external dimension (a changing environment), which both require attention (cf. Gonzalez 2012c). In fact, the variability of PGI can be characterised in terms of historicity.[5]

---

[5] A complete study of the role of historicity in the dynamics of science and the evolution/revolution of associated concepts can be found in Gonzalez (2011, 2012c, 2017b).

Both types of complexity are also present in economics and the sciences of the Internet. Economics consists of a complex structure (semantic, logical, epistemological, methodological, ontological, axiological and ethical) and is aimed at explaining and predicting phenomena. On the other hand, economics also has complex dynamics that must solve specific problems within an ever-changing social environment. As an applied science, economics thus combines prediction and prescription. From a general point of view, Rescher (1998) and Simon have paid more attention to structural complexity than to dynamic complexity, even though Simon considers economics as a science of the artificial (cf. Gonzalez 2012a). Recently, several studies have characterised the sciences of the Internet—and the disciplines that that use this network to enlarge their field of action or to create novelties in a strict sense—as applied science, within the realm of the artificial, with specific complexity levels (cf. Gonzalez 2018b; Gonzalez and Arrojo 2019).

This complexity in the structural and dynamic dimensions in PGI affects the success of new designs and makes PGI an applied science, within the realm of the artificial, where scientific prediction is an aspect of great interest.

9.2 Scientific Prediction in Plant Genetic Improvement as an Applied Science of Design

Prediction is among the central topics in twentieth century and early twenty-first century philosophy and methodology of science. The act of predicting is involved in the process of inquiry, leading to future knowledge as well as in the investigation of *new facts*. Prediction is particularly important in basic science—oriented towards expanding knowledge—and in applied science—aimed at solving specific problems (cf. Gonzalez 2010b, 2018b). From an axiological point of view, scientific prediction can be understood as the main aim of the science of PGI as an applied science of design: it is essential to understand the potential future in order to make a new design.

In the philosophy of science there is a relevant number of thinkers that highlight the role of scientific prediction. These thinkers either emphasise the role of prediction as an aim, as a test or as a guide for practical action. This field has many supporters, including Francis Bacon, August Comte, Hans Reichenbach, Karl Popper and Imre Lakatos, among others. For these luminaries, prediction is the most important aspect of science. However, other authors like Stephen Toulmin—with an instrumental view of prediction—have understood prediction as a mere skill or ability subordinate to the explanation (cf. Gonzalez 2013c).

From the epistemological and methodological point of view, prediction is a central question in applied science, insofar as it seeks to solve specific problems, which requires an understanding of the possible future. Prediction thus has a practical component. It is the step prior to the application of biological science as an applied science of design in a given context (cf. Gonzalez 2015a). In general, prediction requires a series of data (which need a context explaining the scale). These data become information by analysing them in a given context and then finally become knowledge (which requires the categorisation of information).

In this context, the new demands of PGI as a design science make scientific prediction even more important within this science. We have to make a prediction of the future with respect to the usefulness and degree of acceptance of the new design. These new designs will be developed over long periods of time. In the case of fruit trees, for example, new designs require around 12 years. For this reason, the prediction is a central question in this new *design* (variety) to guarantee its success.

9.2.1 Predictive Knowledge of the Future and its Impact on Plant Genetic Improvement as an Applied Science of Design

From an axiological point of view, scientific prediction in PGI should be seen as the main goal of this science as a design science. In general, prediction is a central concept of the science of the artificial, as Simon points out (cf. Simon 1996). Prediction consists of the anticipation of something before it is known (epistemological question) or before its existence (ontological question). Prediction can be seen in connection with the different components of science: language, structure, knowledge, method, activity, ends and values (cf. Gonzalez 2010b, 2015b). In this regard, predictive knowledge—the anticipation of the possible future—can be used to evaluate the progress of scientific activity. It is very attractive from the scientific point of view in natural sciences, social sciences and the sciences of the artificial.

In the case of PGI as an applied science of design, prediction is a key factor. Successful PGI is associated with the success of the prediction in the adaptability and acceptance of the new design or variety. This is similar to that which occurs in other design sciences such as pharmacology. In the case of the genetic improvement of fruits, predictive knowledge is of special interest because, as mentioned above, the *design* (the new variety) takes between 10 and 15 years to be developed (cf. Martínez-Gómez et al. 2003). This perspective of prediction in applied science is different from Hans Reichenbach's concept of scientific prediction in the case of basic science (cf. Reichenbach 1938). In his book *Experience and Prediction*, this author focusses his study of prediction on basic science. In basic science, Reichenbach describes that the basis for a successful prediction lies in induction and the theory of knowledge.

9.2.2   Levels of Predictive Knowledge in Plant Genetic Improvement as an Empirical and Applied Science of Design

In empirical sciences, such as PGI, the scientific processes can work on different levels of reality (micro, meso or macro). This is a more or less holistic perspective of the process. The different levels of reality in PGI are associated with molecular biology (micro level), the genetic constitution of the individual (meso level) and the phenotype or overall appearance of the individual (macro level).

Macro level: Since more than 10,000 years ago, humans have seen that it is possible to select plants to facilitate cultivation for food production, which is what we call *agriculture*. The selection process is called adaptation and domestication, and it involves a series of practices called the *Genetic Improvement of plants*. Since early times, PGI has been a kind of empirical knowledge, based on trial and error, without the reliability of what we now consider a science.

During the first millennia BC, the procedures used in agriculture were relatively poor and largely based on observation. Later, during the first century BC, the Hispanic-Latin writer Columella wrote in his *De re rustica*, "if the crop is exceptional, the seed grains should be moved around in a bowl with water, and those that sink to the bottom should be used for reproduction" (Buiatti 2004, 5). In this first stage (pre-genetic), hybridisation and crossbreeding in plants were disciplines of applied interest, exclusively used within botany. Yet progressively these disciplines acquired more rigour in the process of solving specific problems. Plant breeders observed that the characteristics related to increasing quantity (production, yield, grain size, etc.) were of particular interest in improvement efforts. To improve meant to move *towards more* in the overall improvement of a whole organism, such as a plant. This macro level of knowledge is still the level at which PGI is carried out.

Meso level: Since the publication of the work of Gregor Mendel in 1866, more accurate methods have been used to explain the transmission of characteristics within PGI (Artola and Sánchez-Ron 2012). In this regard, the three following developments are the most important contributions to the biological sciences in the twentieth century: (i) the discovery of the DNA structure (1953); (ii) the definition of the gene at the molecular level; and (iii) the development of the central dogma of molecular biology (1970) (genetic information may be transferred between the nucleic acids to proteins, including DNA replication, RNA transcription and translation into protein expressed in the phenotype) (cf. Solís and Selles 2009). These discoveries about the molecules involved in inheritance were incorporated into the methodologies applied in biology and PGI in the so-called *Genomic* era (Watson 2004). We can thus consider an intermediate level (meso level) between the whole organism (macro) and the DNA (micro level).

Micro level: Subsequently, as a result of the development of DNA sequencing techniques, it became possible to access genetic knowledge at the level of each nucleotide in the so-called *Genomics* era. At present, in the *Post-genomic* era, we have a new scientific revolution in PGI. This so-called *Post-genomic era* is characterised by three elements that can cause a *paradigm shift* in the existing approaches: a) the incorporation of new methods of high-throughput sequencing of both DNA and RNA; b) the development of complete reference genomes of different species; and c) the change in perspective on the expression of traits (based on the project ENCODE, The Encyclopaedia of DNA Elements), where the centre of gravity of these processes is placed in the study of RNA rather than DNA (Martínez-Gómez et al. 2012). Each of these stages has contributed methodological developments that are applicable to PGI (cf. Hayward et al. 1993).

Following the historical analyses of the evolution of knowledge in PGI, we have gone from a macro level to a meso level and, finally, to a micro level. This methodological variety is open to a number of consequences, such as differences in the research according to levels of reality (micro, meso and macro). This leads to multiscale modelling and to questioning the "fundamental" parts in PGI, understood as the necessary support for the whole discipline; the need to assess the efficacy of procedures and methods of scientific activity in engendering high quality results in research; the relevance of contextual factors for PGI methodology; the recognition of a plurality of stratagems when doing research in PGI as an empirical science; and the need for an ethical component while developing scientific methods, because values should have a role in scientific research.

9.2.3 Types of Prediction and their Presence in Plant Genetic Improvement as an Applied Science of Design

Quantitative prediction has the highest scientific value for certain authors, such as Thomas Kuhn (Kuhn 1970). This type of prediction is the most prestigious from the scientific point of view, but it is difficult to implement in developing new designs, which is the case in PGI. Quantitative prediction is the most rigorous type of prediction and, in many cases, it is based on accurate mathematical models (cf. Gonzalez 2008).

Within quantitative prediction, we should specify the concept of *precision* or level of information regarding details, and the concept of *accuracy* or correctness in the statement made about the future. Quantitative prediction requires a theoretical model. In PGI, this type of prediction is possible at micro levels and to a lesser extent at the meso level. Quantitative predictions are those made to determine the function of DNA and RNA (micro level) or in performing a statistical analysis of the inheritance of genes (meso level) (cf. Martínez-Gómez et al. 2012).

On the other hand, we can consider qualitative prediction as a kind of explanatory prediction with less rigour or *prestige* from the scientific point of view. Qualitative prediction plays a prominent role in the process of methodologically contrasting in any applied science of design. Within the realm of the social sciences, the most striking expression of this process of contrasting has been the methodological controversy between *explanation* (*Erklären*) and *understanding* (*Verstehen*).[6] In the area of economics, the debate between predictions versus understanding is on the same level.[7]

The use of qualitative predictions is of great interest in PGI, especially at the macro level, since such predictions help to produce new designs. In the case of plant improvement, the figure of the expert is very important in this type of prediction. Intuition also plays an important role when something is predicted without a mediated process. This is the case of prediction in PGI as a design science, which includes additional complexity that goes beyond the field of nature itself.

9.2.4   The Role of Prediction in the Objectives of Design: The Need for Prediction for Prescription

In order to prescribe how to solve problems, we need predictions of future possibilities. But prediction runs into the limits of science. The limits of prediction can thus be seen as the limits of science, which can be also seen as *barriers* of prediction.[8] This leads to the fact that something may not be predictable at this time (we do not currently have models) or may be entirely unpredictable (we never will have models) (cf. Gonzalez 2010b). Within PGI, the perspective of the limits of prediction is very important.

Most of the studies on the prediction required for prescription have been conducted in the field of economics (cf. Friedman 1974; Gonzalez 1998a, 2015b), but more recent studies have been carried out in biological sciences (cf. Gonzalez 2015a) and the sciences of the Internet (cf. Gonzalez 2015a, 2018b). The group of Michael Friedman is focussed on economic theory and sees prediction as fundamental at the methodological level in basic science. Such researchers emphasise the use of prediction as a scientific test of discipline. Meanwhile, we have a weak line of researchers such as Herbert Simon who claim that making predictions is not a priority in science.[9] For Simon, it seems more important to understand processes, especially for making decisions.

In PGI, particularly in the case of obtaining new fruit varieties, the need for prescription based on prediction is imperative in the context of the design sciences, given the aforementioned fact that designs (new varieties) must be planned out

---

[6]Different approaches to this methodological controversy can be found in Gonzalez (2015c, 173–177).

[7]This second question guides us to the methodological preference of Alexander Rosenberg, who remarks on the role of prediction in contraposition to persuasion, see Rosenberg (1992).

[8]We can distinguish between the limits or barriers of prediction and the confines or ceiling, see Gonzalez (2015b, 2016).

[9]For a detailed description see Simon (1996).

12 years in advance, which is the average time required to develop a new fruit variety (cf. Martínez-Gómez et al. 2003).

In this case, prescription is aimed at making decisions in the process of PGI. The choice of the progenitors to be used is the key in this process. Performing crosses is the basis of the design. These crosses can be of the complementary type, when we cross two varieties with complementary characteristics to obtain a new variety that combines the good traits of both varieties, or of the transgressive type, where two varieties with good traits are crossed in order to obtain an even better variety. From an economic point of view, prescription is intended to select the number of hybrids to be obtained in each crossing. This number is directly related to the cost of the evaluation process of the hybrid, which is an important part of the overall cost of the process.

9.3 Methodological Orientations in Scientific Prediction in Plant Genetic Improvement as an Applied Science of Design

Regarding the methodological orientations within the realm of prediction, we can consider three methodological levels. There is one general level, where methodology could encompass, in principle, all empirical sciences; there is a second special and more limited level, which deals with groups of sciences (natural, social or artificial); and then there is a third level, which is specific to a particular scientific discipline.

These methodological orientations are also associated with the level of reality in the case of PGI. The broader methodologies are based on the level of molecular biology (micro level) and the genetic constitution of an individual (meso level). However, at the macro level, the methodology used is more restricted and only applicable to PGI as a design science.

9.3.1 Levels and Reliability of Prediction

Concerning the levels of prediction, in the case of physical models, which are usually much simpler than social models, quantitative prediction is presented as exact models (cf. Friedman 1974). In our case, in biology, genetic individuals have an intermediate level of complexity that lies between physical entities and societies. As for reliability, Reichenbach argued that absolute reliability cannot be guaranteed in predictions (cf. Reichenbach 1938). The reliability of the prediction is determined by whether there is a causal component. This approach is more realistic in the case of PGI. Moreover, in our case, the event occurs within the world of the design science, where the human factor can also have an effect.

Furthermore, the number of variables known and the level of knowledge that we have about them determines the quality of the prediction. In this context, Wenceslao J. Gonzalez has proposed differentiating between types of predictions, according to the degree of control of the variables analysed. The concepts are foresight, prediction and forecasting, together with the concept of planning. They reveal new philosophical and methodological possibilities regarding the anticipation of the possible future (cf. Gonzalez 2010b, 2015b).

When something is a foresight, there is clear control of the variables involved in the statement regarding the future. Thus, there is more certainty than in the other two cases. Following a foresight, it is therefore possible to anticipate the future well enough, which facilitates making a good prescription. In the case of prediction, some of the variables are not actually under our control. The anticipation of the future is thus made with some degree of reliability, which is less than in the previous case.[10]

In terms of forecasting, there is no clear control of the variables, so the anticipation of the future is less reliable than in the case of foresight or prediction. Forecasting includes a certain margin of error as part of the content regarding the future, which is also the case when we have random variables. Furthermore, we can also talk about planning phenomena as the goal of the decision-making process after making a prediction. This planning is a decision made for the short, medium or long term. Such planning can be based on a level of uncertainty, depending on the kind of phenomenon studied.

9.3.2   The Evolution of Plant Genetic Improvement: Historicity in Prediction and the Impact of Historicity on Methodological Approaches

Scientific methods are a means to develop knowledge and to evaluate it in terms of progress. Change may lead to progress and the analysis of that change must contemplate historicity (cf. Lakatos 1978). For researchers like Imre Lakatos, if a research programme does not progress, it is a programme that, in fact, degenerates. In science, we can therefore distinguish between the history of what happened (which is expressed in data) and the theory (the formal diagrams that explain or organise) (cf. Mosterín 2013). In this perspective, it seems clear that there has been progress in the history of PGI. But there is also historicity in the content of biology, in general, and genetics, in particular, which follows its own design.

To recognise the historicity of human actions in our case, we must look to the history of the evolution of PGI. This means having knowledge of the previously developed designs (the new varieties) in order to better understand the behaviour of

---

[10]For a detailed description of these concepts, consult Gonzalez (2010b) and Gonzalez (2015b, 251–284).

these new designs in the future. Even though we may be merely projecting into the future, taking past designs into account nevertheless offers an approach to making scientific predictions in the field of PGI. The role of novelty is relevant here: every year it is necessary to review the current approaches to design in light of the new data available. This revision should be influenced by the success of other designs already in use, progress in our understanding of the genetic bases of our work and the availability of new methodological approaches.

9.3.3   Internal and External Variables to Consider in Making Predictions in Plant Genetic Improvement

To develop good predictive models, we need to know the relevant variables that influence our knowledge of the possible future in the field studied. It is thus necessary to examine the endogenous (internal) and exogenous (external) variables influencing the phenomenon in question. Both types of variable must be subjected to scientific evaluation. The first internal variables to consider in any PGI programme are derived from plant traits that are considered as objectives. Using fruit breeding as an example, the most important characteristics in breeding programmes are as follows (Martínez-Gómez et al. 2003):

- Tree: Floral self-compatibility, time of flowering and maturation, productivity and resistance to pests and diseases.
- Fruit/seed (almond): Organoleptic quality of the fruit, fruit and seed size, fruit and seed shape, seed taste, hardness of the shell and seed bitterness.

Internal variables also include the current methodologies available for use in the selection of individuals. These methodologies are closely related to the level of knowledge available at the meso and micro levels.

In addition, we can include various economic factors within the production framework, i.e., the aims, processes and results. These factors affect the external economic variables that have an impact on market values. This is because the result of a new variety (the *bio-artefact*) has a certain durability (life cycle) and a market price.

In the genetic improvement of fruit species, it is necessary to consider a number of external variables, including interactions with the environment over a period of several years. There are many characteristics, such as flowering time and floral compatibility, that must be evaluated for at least 3 years, when trees produce the first flowers. Due to such drawbacks, the use of molecular marker-assisted selection methods is of particular interest in breeding programmes (cf. Martínez-Gómez et al. 2003).

Satisfying consumer demands is also an important external variable (cf. Infante et al. 2011). New designs must have the appropriate characteristics to meet such demands. In addition, the availability of different techniques for analysis and selection is another important external variable. Indeed, PGI is defined as "the application of genetic techniques to develop new plant varieties that surpass existing varieties in terms of productivity, quality, or durability" (Hayward et al. 1993, p. 5).

The degree of knowledge of these internal and external variables determines the quality of the prediction in the design of the new plant varieties developed, in addition to its effectiveness, suitability and feasibility.

Finally, together with knowledge, which also has economic value, making predictions in PGI also involves human activity that is geared towards reaching feasible goals in the field. The processes to be performed must have a reasonable cost in terms of the level of effort required to carry out the design. The result is the production of a new bio-artefact, which requires scientific knowledge and scientific processes. But the real impact on society is produced by the technology that fosters such innovation resulting from objectives and work. Overall it can be said that economics directly affects the human activity performed (cf. Gonzalez 1998b). Thus, scientific rationality, technological rationality and economic rationality intervene (Gonzalez 1998b). Economic mediation directly affects the human activity deployed (cf. Gonzalez 2001).

9.3.4 Review of Scientific Predictions as a Form of Self-Correction

All relevant internal and external variables must be considered for prediction. In addition, there must be a capacity for self-correction and change. Reichenbach argued that predictions are essentially correctable (cf. Reichenbach 1938). And, certainly, some predictions are better than others, because they are right in a greater number of cases (cf. Gonzalez 2010a).

While science can be considered as a revisable human activity, it is also true that there are different levels of reliability within available knowledge. Popper defended a critical attitude in scientific and rational activity (cf. Popper 1959). PGI as a design science must be undertaken with such a critical attitude so that there is a continuous process of review and self-correction of the predictions made in order to elaborate appropriate prescriptions. Every year it is necessary to predict the desired results and prescribe the types and numbers of crossings that are to be performed.

9.4 Concluding Remarks

Based on the analysis made in this paper we can draw the following conclusions: (1) PGI is an applied science, because it is aimed at solving specific problems within a given setting. (2) PGI includes two sides: it is both an empirical science that deals with natural phenomena (natural science) and also a science that makes designs that expand human possibilities (a science of the artificial). (3) The emphasis on PGI as a human activity rather than mere content shows the advantages of pragmatism and pluralism as methodological prospects in scientific research. (4) PGI as an applied science of design is complex, not only on the ontological and epistemological levels, but also in terms of its structural and dynamic complexity. (5) In PGI, an empirical science that is articulated as a design science, there are three levels of reality: (i) molecular biology (micro level); (ii) the genetic constitution of an individual (meso level); and (iii) the phenotype or overall appearance of the individual (macro level). (6) The type of prediction and the methodology to be used on these three levels depends on the levels of reality described here. (7) Prediction is a key aim of this design science and is needed in order to open the door to prescription. (8) The historicity of human activity in any design is a criterion of analysis for the future since it affects the predictions made in PGI. (9) Finally, it is also necessary to consider several different variables, both internal and external, when making predictions and prescriptions.

References

Artola, M., & Sánchez-Ron, J. M. (2012). *Los pilares de la Ciencia*. Madrid: Espasa.

Buiatti, M. (2004). *Le Biotecnologie*. Roma: Il Mulino.

Cuevas-Badallo, A. (2005). A model-based approach to technological theories. *Techné, 9*, 18–49.

Cuevas-Badallo, A. (2008). Los bio-artefactos: viejas realidades que plantean nuevos problemas en la adscripción funcional. *Argumentos de Razón Técnica, 11*, 71–96.

Friedman, M. (1974). Explanation and scientific understanding. *The Journal of Philosophy, 71*, 5–19.

Gonzalez, W. J. (1998a). Prediction and prescription in economics: A philosophical and methodological approach. *Theoria, 13*, 321–345.

Gonzalez, W. J. (1998b). Racionalidad científica y Racionalidad tecnológica: La mediación de la racionalidad económica. *AGORA-Papeles de Filosofía, 17*, 95–115.

Gonzalez, W. J. (2001). De la Ciencia de la Economía a la Economía de la Ciencia: Marco conceptual de la reflexión metodológica y axiológica. In A. Ávila, W. J. Gonzalez, & G. Marqués (Eds.), *Ciencia económica y Economía de la Ciencia: Reflexiones filosófico metodológicas* (pp. 11–37). Madrid: FCE.

Gonzalez, W. J. (2002). Caracterización de la 'explicación científica' y tipos de explicaciones científicas. In W. J. Gonzalez (Ed.), *Diversidad de la explicación científica* (pp. 13–49). Barcelona: Ariel.

Gonzalez, W. J. (2005). The Philosophical Approach to Science, Technology and Society. In W. J. Gonzalez (Ed.), *Science, technology and society: A philosophical perspective* (pp. 3–49). A Coruña: Netbiblo.

Gonzalez, W. J. (2007). Análisis de las Ciencias de Diseño desde la racionalidad limitada, la predicción y la prescripción. In W. J. Gonzalez (Ed.), *Las Ciencias de Diseño: Racionalidad limitada, predicción y prescripción* (pp. 3–43). A Coruña: Netbiblo.

Gonzalez, W. J. (2008). Rationality and prediction in the science of the artificial: Economics as a design science. In M. C. Galavotti, R. Scazzieri, & P. Suppes (Eds.), *Reasoning, rationality, and probability* (pp. 165–186). Stanford: CSLI Publications.

Gonzalez, W. J. (2010a). Recent approaches on observation and experimentation: A philosophical-methodological viewpoint. In W. J. Gonzalez (Ed.), *Methodological perspectives on observation and experimentation in science* (pp. 9–48). A Coruña: Netbiblo.

Gonzalez, W. J. (2010b). *La predicción científica: Concepciones filosófico-metodológicas desde H. Reichenbach a N. Rescher.* Barcelona: Montesinos.

Gonzalez, W. J. (2011). Conceptual changes and scientific diversity: The role of historicity. In W. J. Gonzalez (Ed.), *Conceptual revolutions: From cognitive science to medicine* (pp. 39–62). A Coruña: Netbiblo.

Gonzalez, W. J. (2012a). La Economía en cuanto a Ciencia: Enfoque desde la complejidad. *Revista Galega de Eonomía, 21*(1), 1–30.

Gonzalez, W. J. (2012b). Complejidad estructural en Ciencias de Diseño y su incidencia en la predicción científica: El papel de la sobriedad de factores (*parsimonious factors*). In W. J. González (Ed.), *Las Ciencias de la Complejidad: Vertiente dinámica de las Ciencias de Diseño y sobriedad de factores* (pp. 143–167). A Coruña: Netbiblo.

Gonzalez, W. J. (2012c). La vertiente dinámica de las Ciencias de la Complejidad. Repercusión de la historicidad para la predicción científica en las Ciencias de Diseño. In W. J. Gonzalez (Ed.), *Las Ciencias de la Complejidad: Vertiente dinámica de las Ciencias de Diseño y sobriedad de factores* (pp. 73–106). A Coruña: Netbiblo.

Gonzalez, W. J. (2013a). The science of design as science of complexity: The dynamic trait. In H. Andersen, D. Dieks, W. J. Gonzalez, T. Uebel, & G. Wheeler (Eds.), *New challenges to philosophy of science* (pp. 299–311). Dordrecht: Springer.

Gonzalez, W. J. (2013b). The roles of scientific creativity and technological innovation in the context of complexity of science. In W. J. Gonzalez (Ed.), *Creativity, innovation, and complexity in science* (pp. 11–40). A Coruña: Netbiblo.

Gonzalez, W. J. (2013c). Scientific prediction in the beginning of the historical turn: Stephen Toulmin and Thomas Kuhn. *Open Journal of Philosophy, 3*(2), 351–357.

Gonzalez, W. J. (2015a). Prediction and prescription in biological systems: The role of Technology for Measurement and Transformation. In M. Bertolaso (Ed.), *The future of scientific practice: 'Bio-Techno-Logos'* (pp. 133–146). London: Pickering and Chatto.

Gonzalez, W. J. (Ed.). (2015b). *Philosophico-methodological analysis of prediction and its role in economics.* Dordrecht: Springer.

Gonzalez, W. J. (2015c). From the characterization of 'European philosophy of science' to the case of the philosophy of the social sciences. *International Studies in the Philosophy of Science, 29*(2), 167–188.

Gonzalez, W. J. (2016). Rethinking the limits of science: From the difficulties to the Frontiers to the concern about the confines. In W. J. Gonzalez (Ed.), *The limits of science: An analysis from "Barriers" to "Confines"* (pp. 3–30). Leiden: Poznan Studies in the Philosophy of the Sciences and the Humanities/Brill-Rodopi.

Gonzalez, W. J. (2017a). From intelligence to rationality of minds and Machines in Contemporary Society: The sciences of design and the role of information. *Minds and Machines, 27*(3), 397–424.

Gonzalez, W. J. (2017b). Cambio conceptual y diversidad científica: El papel de la historicidad en la dinámica de la Ciencia. *Factótum. Revista de Filosofía, 18*, 10–32.

Gonzalez, W. J. (2018a). Complejidad dinámica en Internet como plataforma de información y comunicación: Análisis filosófico desde la perspectiva de Ciencias de Diseño y el papel de la predicción. *Informação e Sociedade: Estudos, 28*(1), 155–168.

Gonzalez, W. J. (2018b). Internet en su vertiente científica: Predicción y prescripción ante la complejidad. *Art, 7*, 75–97.

Gonzalez, W. J., & Arrojo, M. J. (2019). Complexity in the sciences of the internet and its relation to communication sciences. *Empedocles: European Journal for the Philosophy of Communication, 10*(1), 15–33.

Hayward, M. D., Bosemark, N. O., & Romagosa, I. (1993). *Plant breeding. Principles and prospects.* London: Chapman and Hall.

Infante, R., Martínez-Gómez, P., & Predieri, S. (2011). Breeding for Fruit Quality in *Prunus*. In M. A. Jenks (Ed.), *Breeding for fruit quality* (pp. 201–229). New York: Blackwell.

Jasanoff, S. (2006). Biotechnology and empire: The global power of seeds and science. *Osiris, 21*, 273–292.

Kuhn, T. S. (1970). *The structure of scientific revolutions* (2nd ed.). Chicago: University of Chicago Press.

Lakatos, I. (1978). *The methodology of scientific research programs, philosophical papers.* Cambridge: Cambridge University Press.

Martínez-Gómez, P., Sozzi, G. O., Sánchez-Pérez, R., Rubio, M., & Gradziel, T. M. (2003). New approaches to *Prunus* tree crop breeding. *Journal of Food Agriculture and Environment, 1,* 52–63.

Martínez-Gómez, P., Sánchez-Pérez, R., & Rubio, M. (2012). Clarifying Omics concepts, challenges and opportunities for *Prunus* breeding. *OMICS, 16,* 268–283.

Mendel, G. (1866). Versuche über Pflanzen-Hybriden. *Verhandlungen des Naturforschenden Vereines, Abhandlungen, Brünn, 4,* 3–37.

Mosterín, J. (2013). *Ciencia, Filosofía y Racionalidad.* Barcelona: Gedisa.

Niiniluoto, I. (1993). The aim and structure of applied research. *Erkenntnis, 38,* 1–21.

Popper, K. (1959). *The logic of scientific discovery.* London: Routledge Classics.

Reichenbach, H. (1938). *Experience and prediction.* Chicago: The University of Chicago Press.

Rescher, N. (1998). *Complexity: A philosophical overview.* New Brunswick: N. J. Transaction Publishers.

Rosenberg, A. (1992). *Economics–mathematical politics or science of diminishing returns?* Chicago: The University of Chicago Press.

Simon, H. (1996). *The science of the artificial* (3rd ed.). Cambridge, MA: The MIT Press.

Solís, C., & Selles, M. (2009). *Historia de la Ciencia.* Madrid: Espasa.

Watson, J. D. (2004). *DNA. The Secret of Life.* New York: Alfred A. Knoff.

Part V
Methodological Pluralism in Social Sciences and Ethical Values

Part V
Methodological Pluralism in Social
Science and Ethical Values

Chapter 10
Challenges to Validity from the Standpoint of Methodological Pluralism: The Case of Survey Research in Economics

María Caamaño-Alegre

Abstract By focusing on the case of survey research in economics, the paper shows how methodological pluralism emerges as a natural consequence from a very common dynamics of feedback between problems and solutions taking place in scientific practice. This continuous feedback between methodological problems and attempts at solving them, being essentially connected with the pursuit of validity, naturally leads to the pluralistic tendency found in empirical research in economics over the last decades and clearly manifest in the case of survey research. The methodological challenges within the latter mainly come from the pervading presence of framing effects in survey research, which, as argued here, prompts the application of new procedures able to improve the different kinds of validity.

Keywords Validity · Survey research · Framing effects · Methodological pluralism

The idea that science makes progress from old problems to new problems underlies methodological approaches to science as influential and diverse as those by R. K. Popper (1991/1999, chapter IX), T. S. Kuhn ([1962]1970, chapters IV, IX) or L. Laudan (1981, chapter VII). In traditional, contemporary philosophy of science, much attention has been paid to theoretical problems and questions on the relationship between theory and experience. It has been only relatively recently, over the last decades, that there has been a turn towards the study of scientific

This work has been financially supported by the research projects "Laws and Models in Physical, Chemical, Biological, and Social Sciences" (PICT-2018-03454, ANPCyT, Argentina), and "Stochastic Representations in the Natural Sciences: Conceptual Foundations and Applications (STOCREP)" (PGC2018-099423-B-I00, Spanish Ministry of Science, Innovation and Universities).

M. Caamaño-Alegre (✉)
Dpto. Filosofía, Facultad de Filosofía y Letras, Valladolid, Spain

© Springer Nature Switzerland AG 2020
W. J. Gonzalez (ed.), *Methodological Prospects for Scientific Research*, Synthese Library 430, https://doi.org/10.1007/978-3-030-52500-2_10

Table 10.1 Feedback dynamics between problems and solutions leading to methodological pluralism

	First stage	Second stage	Third stage	Following stages
Feedback between problems and solutions	Empirical Problems associated to procedure 1: $Prob_1$	Solutions to $Prob_1$ dependent on procedure 2: $Proc_2$	Problems associated to $Proc_2$: $Prob_2$	Feedback between problems $Prob_2$, ... $Prob_j$ and solutions $Proc_3$, ... $Proc_n$ continues
Pluralistic tendency	Employment of procedure 1 ($Proc_1$)	Employment of $Proc_1$ y $Proc_2$	Employment of $Proc_1$ y $Proc_2$	Employment of $Proc_1$, ... $Proc_n$

practice, opening the possibility to explore the *praxical* and experimental side of scientific development. As it turns out, however, the inquiry into scientific practice reinforces the view that science is a problem-solving activity, one in which methodological refinement and innovation plays a major role. The abovementioned process of feedback between problems and attempts at solving them very often requires the devise of new methodological options, thereby favoring the pragmatic vindication of a certain methodological pluralism. In what follows I analyze how this dynamics unfolds within the field of empirical research in economics, particularly in the subfield of survey research. The schematic features of such dynamics are represented in Table 10.1 appearing above.

Over the last decades, serious empirical limitations associated to traditional observational methods in economics has led economic researchers to a methodological pluralism, which in turn has given rise to new problems as well as new attempts at solutions. More broadly, in the case of social sciences, where intervention in the subjects' behavior is frequently verbal or at least language-dependent, the use of a wide variety of linguistic means to gather information about their beliefs, expectations, assessments or planned courses of action has significantly increased. In this respect, the reliance on various kinds of surveys and interviews has extended substantially, thereby widening the scope of linguistic interventions beyond the directions verbally conveyed to the experimental subject. In economics, observation of choices was supplemented by survey research to overcome the ambiguity of these observations, hence improving the validity of causal inferences about preferences. Yet, survey research faces its own validity challenges in the form of framing effects, which threatens all the main forms of validity as well as the related feature of reliability. Consistently divergent answers to apparently the same questions concerning preferences preclude any possibility of determining robust correlations between questions regarding options and answers expressing preferences, making thus impossible to draw sound causal inferences and ultimately to attain further forms of validity that presuppose reliability, robust correlations and sound causal inferences.

The discussion of the above issues is structured as follows: first, the standard notion of validity in research methodology is characterized; second, the methodological pluralism connected with the increasing use of linguistic interventions within economic methodology is examined; third, the challenge of detecting, explaining and controlling framing effects is analyzed; finally, some conclusions as to the validity challenges involved are drawn from the previous discussion.

10.1 The Standard Characterization of Validity in Research Methodology

Traditional philosophical approaches to validity were especially concerned with both theory testing and the attribution of logical rationality to science (see Messick's 1989 discussion of the subject). The two classical accounts of validity in Philosophy of Science, putting aside the more recent contributions by philosophers of experiment like I. Hacking (1983), A. Franklin (2005), P. Galison (1997), F. Steinle (1997) or D. G. Mayo (1996), respectively revolve around the notions of verifiability and falsifiability. A more extensive discussion of experimental validity was attempted instead within the emergent field of social science, specifically, within the methodology of empirical psychology. In this field such discussion embraces more than just those research components devised for the purpose of testing a theory. Here we pay attention to the enlarged view on validation coming from the social sciences. Within this area, validity and reliability are characterized both as logically independent notions and as commonly associated properties of measurements and procedures. Reliability concerns the extent to which an experiment, test, or any measuring procedure yields the same results in repeated trials under the same conditions (Pelham and Blanton 2003, 70–77, Carmines and Zeller 1979, 11–13), while the validity concerns the degree of success in attaining the purported outcome (that is, in determining the variable under study). The common association between reliability and validity is due to the fact that the first is usually required in order to establish the validity of the procedure or just to guarantee its useful applicability. Pelham and Blanton (2003, 70–75) distinguish three main forms of reliability: inter-observer agreement, internal consistency (or inter-item agreement in the same test), and temporal consistency (or test-retest reliability). A variety of statistical methods have been developed for the latter's careful assessment. The unreliability of a method has two possible, general sources: the uncertainty of the phenomena measured, and the errors of measurement (whether chance error, systematic or instrumental).

The notion of validity was originally developed from two different traditions in social science, namely, experimental and test research (Table 10.1). In the 1950s, the basic distinction corresponding to the first tradition was that between internal and external validity (Campbell 1957). As for the second tradition, the main kinds of validity were criterion, content, and construct validity (Cronbach and Meehl 1955). In an attempt to cope with different methodological challenges, test community gradually embraced an enlarged and unitary concept of validity (Angoff 1988, 25;

Sireci 2009), one based on a comprehensive notion of construct validity which comprises all sorts of empirical support for test interpretation and use (Messick 1989). Contrary to this, Shadish et al. (2002) kept the primary association of validity with the truth of knowledge claims, and integrated criterion, internal, construct, and external validity in other unitary framework applicable to either kind of social empirical research (i.e., experimental and test-based). Let us briefly consider each of these kinds of validity.

Statistical conclusion validity (or criterion-related validity as labeled by Carmines and Zeller 1979),[1] in the case of two variables, concerns the appropriate use of statistics to infer whether the presumed independent and dependent variables are correlated. It thus refers to how large and reliable is the co-variation between the presumed cause and effect (Campbell 1986).[2] That which is targeted by the measurement constitutes the criterion-variable to assess that validity of the method (or instrument). To establish criterion-related validity it is necessary to measure how well one variable (or set of variables), usually called "independent variable" or "intermediate variable", predicts an outcome, usually called "dependent variable" or "ultimate variable", based on information from other variables. Criteria validity depends on the extent to which the measures are demonstrably related to concrete criteria in the "real" world. When the criterion variable has current rather than future existence, the validity involved is called "concurrent validity", otherwise, this kind of validity is referred to as "predictive validity".

Internal validity refers to whether the co-variation between the presumed independent and dependent variable results from a causal relationship. It is then concerned with the causal interpretation of the criterion-related or statistical conclusion validity. Pelham and Blanton (2003, 62–64) point out that laboratory experiments prove to be very useful in providing information about causality, since they make it possible to isolate independent variables from potential sources of contamination, thereby providing better conditions for controlling individual differences.

In its standard presentation, construct validity is equated with the evidential basis of test interpretation (Messick 1989, 34). It concerns the extent to which a particular empirical indicator (or a set of indicators) represents a given theoretical concept, that is, the extent to which independent and dependent variables truly represent the abstract, hypothetical variables of interest to the scientist (Pelham and Blanton 2003, 66; Shadish et al. 2002, 65). The evaluation of construct validity involves close examination of the auxiliary theory or theories specifying the relationship between concepts and indicators. Such evaluation, therefore, entails examining whether a measure of a construct relates to other measures as established by sound auxiliary hypotheses concerning the construct's empirical content (cf. Carmines and Zeller 1979, 23).

---

[1]Carmines and Zeller have not been included in the above graphical representation because the notion of criterion-validity, even if named differently, had been introduced much earlier.

[2]We should insist here on Mayo's contribution to this topic within the field of philosophy and her emphasis on the significance of statistics for the epistemology of experiment.

External validity refers to the appropriateness of generalizations from results obtained in an experimental setting to phenomena out of such setting. It thus concerns the extent to which a set of research findings provide an accurate description of what typically happens in the real world (Pelham and Blanton 2003, 64). If that which is generalized is a causal relationship, then construct validity consists in the validity of inferences about whether the cause-effect relationship holds over variation in samples, settings, and measurement variables. The two main sorts of generalizations pursued within experimental research are those with respect to some type of entity, and those with respect to some types of situations. On the other hand, the main restrictions to the generalizability of a finding are given by the boundary conditions restricting the attainability of these findings.

One way to address the problem of validation in qualitative terms is to analyze the factors jeopardizing the different kinds of validity. The capability of a method to avoid being affected by those factors can be considered as a sound indicator of how valid the method is. Noise and confounds are the main general threats to both statistical conclusion and internal validity, which are difficult to obtain even in experimental settings where variables can be partially isolated to test their impact on a single dependent variable. On the other hand, construct underrepresentation and construct-irrelevant variance in the test must be emphasized as the general types of threats faced by construct validity. In evaluating the adequacy of con-struct measurement, not only random errors but also systematic ones need to be considered. The latter may occur because of one or both of the following reasons: (*i*) tests leave out something that should be included (according to the construct theory), and (*ii*) they include something that should be left out. There are two main requirements of construct validity needed for protecting interpretations from these general types of threats, namely, convergent and discriminant evidence. The first enables us to assess the degree to which the construct's implications are realized in empirical score relationships, the second provides grounds to argue that these relationships are not attributable instead to distinct alternative constructs (Messick 1989, 34). Finally, some of the typical factors jeopardizing external validity (or generalizability of findings) are the contrived nature of the testing settings, and selection biases. Both issues are related to the general problem concerning the artificiality of laboratory experiments. This problem becomes evident not only in the difficulty to make experimental settings significantly similar to the targeted real situations, but also in the limitations to recruit a group of experimental subjects that are representative enough of the diverse population under study. Pelham and Blanton (2003, 66–67, 176) suggest several ways to minimize the effect of these threats: combining laboratory research with passive observational studies, randomizing subject's selection, and using manipulation checks with the subjects to make sure whether the intended experimental variation corresponds to the one accomplished in the experiment.

10.2   Linguistic Intervention as a Source of Pluralism
in Economic Methodology

From a general point of view, verbal intervention raises two independent problems: the first concerns the possibility of determining whether the meaning assigned by the researcher to the utterances in the research context is the same as that assigned by the respondent; the second is related to the influence that different ways of presenting the same issue may bear on the respondent's response. Even though the focus of this paper is survey research in economics, which is a subject more often connected to the second problem, it may be worth it to mention the recent attempt to address the first issue on the grounds provided by several philosophers of language with crucial contributions within the field of pragmatics. Even if only very tentatively, several possible sources of discrepancy in understanding an utterance have been examined on the basis of John Searle's distinction between linguistic meaning and speaker's meaning – later supplemented by that between linguistic meaning and utterance meaning (cf. Searle 1978). Some of those discrepancies would be caused by the fact that each user of language would associate to an expression some contents beyond its literal meaning, some others would be due to contextual aspects affecting the use of expressions. Paul Grice's pragmatic approach relies on similar distinctions as those drawn by Searle and aims at explaining the same communicative difficulties pointed out by the latter. According to Grice, communicative intentions, which are the essential element in linguistic activity, are subject to the principle of cooperation, which in turn would unfold in different conversational maxims or principles (cf. Grice 1975). The application of this Gricean analysis to the study of empirical research in economics is still at a very early stage (cf. Schwarz 1996; Jones 2007; Geurts 2013), but certainly a pragmatic analysis of the communicative exchanges between the experimenter (or interviewer) and the experimental subject (or respondent) is required in order to determine the discrepancies between the message that the experimenter is trying to convey and the message grasped by the subject.

Linguistic intervention pervades the field of social science, whether by means of directions provided to the experimental subject or by interviews and surveys intended to collect information about the effect that certain issue has on the subjects' beliefs, choices or behavior. In parallel to this variety of verbal interventions, new difficulties concerning the so called "framing effects" are detected, which in turn prompt new attempts at sophisticating the methodological procedures in order to confront them.[3] However, within the sphere of economic methodology, there have been two main prevailing assumptions whose endorsement has led

[3] A traditional methodological principle endorsed in economics, namely, the one establishing that the experimental subject should receive written (not spoken) directions amounts to implicitly acknowledging the risk of introducing unwanted effects and possible confounds within the experimental context through the communicative interaction between the experimenter and the experimental subject.

respectively to either rule out or question the effectiveness of verbal empirical procedures. The first assumption comes from the theory of revealed preference, committed to the methodological principle that inferences about a subject's future choices must be based on observations of previous choices made by the subject. The second assumption, by contrast, is one underlying the economic methodology expanded with survey research and interviews. Such assumption, usually referred to as the principle of extensionality or the invariance principle (Bourgeois-Gironde and Giraud 2009, 385–387), establishes that individuals' preferences should not be affected by variations in the description of a problem. Different ways of presenting the same set of possible options should thus not change the subjects' choices with respect to those options. As we will see next, some of the difficulties related to the above assumptions have motivated the pluralistic expansion of economic methodology.

With respect to the first assumption, it is worth emphasizing that a serious shortcoming affecting the theory of revealed preference stems from the ambiguity of subjects' observable behavior and the resulting inscrutability of expectations or radical under-determination of attribution of expectations. Since expectations, together with preferences, are acknowledged as crucially involved in the subjects' choices, the inscrutability of the former poses a major obstacle to explaining such choices. The problem emerges when researchers need to establish some suppositions about the subjects' expectations in order to make predictions. Given that expectations are not directly accessible through observation, information about them must be obtained by verbal means. The need to validate the verbal means employed in gathering information about expectations emerges from the very recognition that such means are needed in order to make progress. This methodological turn occurring in economics at the beginning of the 1990s results in the recognition of declared preferences, in addition to revealed preferences, as a legitimate evidential source in economics.[4]

As Charles F. Manski points out, from the early 1990s, economists who engaged in survey research have increasingly used questions regarding subjects' probabilistic expectations concerning significant personal events:

> Observed choices may be consistent with many alternative specifications of preferences and expectations, so researchers commonly assume particular sorts of expectations. It would be better to measure expectations in the form called for by modern economic theory; that is, subjective probabilities. Data on expectations can be used to relax or validate assumptions about expectations. Since the early 1990's, economists have increasingly undertaken to elicit from survey respondents probabilistic expectations of significant personal events (Manski 2004, 1329).

Expectations have been determined for various kinds of events, among them: macroeconomic events (stock market returns), the risks faced by a person (job loss,

---

[4]An influential comprehensive criticism of the theory of revealed preference can be found in Daniel Hausman (2012).

Table 10.2 Pluralism in empirical economics emerging from a continuous feedback between methodological problems and attempts at solving them

	First stage	Second stage	Third stage	Forthcoming stage
Feedback between problems and solutions	Problems related to ambiguity of observed behavior: Inscrutability of expectations through observation	Solutions to ambiguity dependent on survey research	Problems associated to declared preference procedures: Framing effects	Solutions to framing effects dependent on self-reports, post-survey questionnaires, mixed representation of outcomes or attributes, mixed modes of representation (both visual and verbal), ...
Pluralistic tendency	Employment of "revealed preferences" procedures	Employment of both revealed and declared preferences procedures	Employment of both revealed and declared preferences procedures	Use of revealed and more refined, robust declared preferences procedures

mortality), future income (earning, Social Security profits), and choices made by a person (purchases, voting choices).

The cycle of methodological refinement described at the beginning of the paper appears here very clearly: the attempt to improve both the predictive effectiveness and the descriptive accuracy of economics goes hand in hand with its methodological extension. As shown in the table below, this widening of the methodological scope leads, like in other fields, to a methodological pluralism of a pragmatic kind (Table 10.2).

The pragmatic side of this tendency towards methodological pluralism should be understood along the lines of the methodological pragmatism put forward by Nicholas Rescher, which closely resembles the one tacitly embraced by researchers in their current practice (cf. Rescher 1977; Suppes 1998). According to this pragmatist standpoint, the question about the validity of procedures is not one to be answered *a priori* (cf. Wiener 1973–1974, 551–556; Haack 2006), but instead one to be assessed according to the usefulness of such procedures to attain certain epistemic ends (cf. Caamaño-Alegre 2013). It must be noted that the essential goal of increasing predictive power will be achieved to the extent that researchers manage to improve statistic, internal, construct and external validity of the procedures they employ. Similarly, descriptive accuracy is closely connected to the validity of the theoretical construct used in explaining behavior. Therefore, the different kinds of validity involve a specification of epistemic ends relative to which understand and evaluate the methodological developments. The growing interest raised by mixed methods and triangulation in economic methodology constitutes another clear sign of the pluralistic tendency in this field, a tendency with the underlying purpose of strengthening both methodological robustness and the empirical adequacy of theories (cf. Dellinger and Leech 2007; Downward and Mearman 2007; Starr 2014; Claveau 2011).

Let us go back to the side of this pluralistic trend that is the focus of this paper, namely, the use of surveys in the context of empirical research in economics. The shortcomings affecting the theory of revealed preference, in particular the need to identify expectations, called for an empirical research by means of surveys, which was initially carried out according to the abovementioned principles of invariance and extensionality. However, despite the use of surveys in economic methodology, it took a long time until the problem of framing effects was properly noticed. Ivan Moscati draws attention to this fact as he states:

> Orthodox economists tend to discard framing effects as manifestations of the irrationality of individuals who simply fail to recognize that identical things are indeed identical. In opposition to this view, Tversky and Kahneman and other behavioral economists have argued that framing effects significantly influence economic behavior and therefore cannot be discarded without weakening the descriptive significance of economic theory; moreover, some framing effects seem to have a rational justification (Moscati 2012, 6–7).

Behavioral economists have therefore diverged from the prevailing view of framing effects in economics, arguing that such effects should be approached, not as mere cognitive flaws in the recognition of identical options, but as signs of the subjects' attitudes towards different aspects involved in those options. So understood, framing effects turn out relevant for the description, explanation and prediction of the subjects' economic behavior. To put it clearly, the methodological problem of framing effects has encouraged the study of the role that language and communication play in subjects' understanding of the described options. As the use of surveys exponentially increases in the economic field, the need to pay attention to framing effects becomes more pressing. Michaela Nardo provides some interesting data in this respect:

> The European Union, as well as the main OECD countries, regularly collect data from business and consumer surveys. The number of these surveys has substantially increased in the last three decades. If in the late 1960s they were less than 30 in 15 countries, in 1997 their number exceeded 300 in 55 countries. Only in the European Union more than 50.000 firms and 20.000 consumers are interviewed each month. Surveys address firms or agents directly, and rather than asking for exact figures, the questionnaires ask for assessment on the movement of short-term variables, such as output, prices, employment, trade, or investments (Nardo 2003, 645).

She warns us, however, that several difficulties underlie the use of (aggregates of results from) surveys with the purpose of building empirically valid representations of expectations as the basis for inferring the agents' future behavior. On the one hand, the agents themselves can fail to estimate their expectations; on the other hand, the frame in which a survey is presented can influence the expectations they declare (cf. Nardo 2003, 657–59).[5] Even though the present paper highlights the

---

[5]Nardo is of course aware of some other possible sources of error not related to the subjects' performance: "If survey data are a poor indicator of agents' expectations, then the quantified proxy will also poorly predict the behavior of the actual economic variable even if agents are

use of surveys for research purposes, it is worth noting that also their practical use entails methodological challenges that need to be addressed in order to guarantee the effectiveness of those procedures as tools for prediction and for gathering of information. In the next section, I examine in more detail what Nardo points out as the second source of difficulties affecting survey research.

10.3 The Problem of Framing Effects: Detection, Explanation and Control

Let us focus now on the specific problem of framing effects and the main attempts at accounting for them. After presenting a comprehensive classification of such effects, I will deal with the current attempts at explaining and controlling them, making special emphasis on the difficulties involved in the pursuit of validity in survey research.

10.3.1 The Detection and Classification of Framing Effects

As soon as the late 1990s, Levin et al. (1998) urged researchers to sophisticate the typology of framing effects so that it became possible to account for the apparently inconsistent results achieved when trying to detect such effects. The plurality of interventions, moreover, entails a corresponding plurality of framing effects whose treatment requires equally differentiated procedures. In the typology suggested by Levin, Schneider and Gaeth, three main kinds of valence framing effects are distinguished: the extensively discussed risky choice framing effect, and two other effects often overseen or mistaken for the latter, namely, attribute framing and goal framing. As explained by the authors (1998, 151, 181), each frame differs from the others in what is framed, what the frame affects, and how the effect is measured.

In the risky choice framing, the complete set of outcomes from a potential choice involving options with different levels of risk is described either in a positive or negative way. The framing effect is here measured comparing the rate of choices for risky options in each frame condition. Risk aversion would explain the fact that, when presented in negative terms, the riskier option is chosen by respondents more often than the safer one. A wide variety of experiments on risky choice,[6]

---

perfectly rational. This is far from trivial, since being survey data approximations of unobservable expectations, they necessarily entail a measurement error. This error can be ascribed to the incorrect scaling of qualitative data, to sampling or aggregation errors and also to the general uncertainty attached to survey figures" (2003, 657).

[6]See Levin et al. (1998, 154–157) for a collection of experimental results obtained within the domain of risky choice framing effects.

from bargain situations to medical treatments, shows that when the outcome is described in terms of gains (lives saved, earned income) subjects' tendency to take risks diminishes. By contrast, such tendency increases when outcomes are expressed in terms of losses (lost lives, incurred debts). The paradigmatic case of risky choice framing effect is illustrated by the so called "Asian disease problem" (cf. Tversky and Kahneman 1981). In this task, the two equivalent pairs of independent options with different level of risk are the following: (a) a sure saving of one-third the lives versus a one-third chance of saving all the lives and a two-thirds chance of saving no lives; (b) a sure loss of two-thirds the lives versus a one-third chance of losing no lives and a two-thirds chance of losing all the lives. The majority of subjects select the first option in the positively framed version of the task, and the second option in the negatively framed version.

In the form of framing called "attribute framing", the positive or negative description of some characteristic of an object or event affects item evaluation, which is estimated by comparing the attractiveness ratings for the single item in each frame condition. The associative processes based on valence is commonly assumed to explain the fact that positively described objects or events are more positively valued. This result has been established with much higher reliability and robustness than the other two kinds of framing effects compared by Levin et al. (1998, 160). The fact that evaluations vary as a result of positive or negative framing manipulation has been established for issues as diverse as consumer products, job placement programs, medical treatments, industry project teams or students' level of achievement or the performance of basketball players.[7] Ground beef, for example, was rated as better tasting and less greasy when it was described as 75% lean rather than as 25% fat. Similarly, students' performance was rated higher when their scores were expressed in terms of percentage correct or percentage incorrect. Analogous results were obtained in the rest of cases.

Finally, in the case of goal framing, the same consequences of a conduct are specified either in positive or negative terms. The positive frame focuses attention on the goal of obtaining the positive consequence (or gain) associated with a given behavior, whereas the negative frame focuses attention on avoiding the negative consequence (or loss) associated with not performing such behavior. The variation in how persuaded an agent is to make or not make the decision to perform a certain conduct is regarded as an effect of the variations in the frames applied. The effect itself is measured by comparing the rate of adoption of such conduct under each frame condition. Experimental evidence shows that the negatively framed message, that is, the one emphasizing avoidable losses, proves more persuasive than the same message framed positively, and therefore stressing the potential gains. Real examples where goal frames are at use can be found in studies on the promotion of health, on endowment or on social dilemmas. Most subjects appear more inclined to adopt a certain conduct, —like for example, breast self-examination, use of

---

[7]See also Levin et al. (1998, 161–163) for a lengthy compilation of experimental results related to attribute framing effects.

public resources or of credit card-, when they receive information stressing the potential losses derived from not engaging in such conduct than when presented with information highlighting the potential profits resulting from engaging in it.

In the abovementioned examples, individuals show themselves more persuaded to adopt a given behavior when descriptions emphasize, respectively, the decrease in the probability of detecting a cancer if there is no self-examination carried out versus the increase of such probability in case a self-examination is performed, the losses suffered by the individual who contributes to the public goods versus the foreseen gains if the individual contributed to them, and the losses due to not using the credit card versus the benefits derived from its use.[8]

10.3.2   The Attempts at Explaining Framing Effects

Despite the growing interest raised by the problem of framing effects, the majority of studies on these effects are focused on their diagnosis, while the attempts at explaining and controlling them are still extremely tentative and fragmentary. As already pointed out by Tversky and Kahneman (1981, 1991) in several of their influential studies on framing effects, the task of devising frames must be done taking into account individuals susceptibility to changes in reference points or in what is perceived as the *status quo* regarding some issue. Different frames would lead to different choices of reference points and, consequently, to a different way to encode the outcomes as gains or losses, which would accordingly bring about a different selection of options. This clearly calls for the development of procedures that can disclose such susceptibility on the side of the respondents.

In developing their prospect theory (cf. Kahneman and Tversky 1979), both authors appeal to the possible occurrence of highly intertwined phenomena like loss aversion and the endowment effect. These phenomena would emerge in most cases due to some framing conditions in which the reference point regarding the value of an outcome does not stay neutral but varies depending on what is induced by the frame itself. In their own words:

> However, the location of the reference point, and the consequent coding of outcomes as gains or losses, can be affected by the formulation of the offered prospects, and by the expectations of the decision maker (Kahneman and Tversky 1979, 274).

Let us recall that prospect theory, as opposed to classical theory, is committed to the view that risk aversion is dependent on a reference point. Under that assumption, it is predicted that risk aversion is linked to the domain of gains and risk seeking to domain of losses. In their paper from 1979, Kahneman and Tversky established that the above tendency could be reversed depending on the framing employed for the same pair of options. An initial remark in that direction can be found in

---

[8]The wide range of real cases collected by Levin et al. (1998) can be found in 169–171.

some of their comments on the isolation effect (Kahneman and Tversky 1979, 271), that is, individuals' inclination to ignore those components shared by alternatives and to focus on those making them different. Since there is more than one way to decompose a pair of alternatives in shared and distinctive components, the different ways of decomposition may also prompt different preferences. This point is made more explicit as both authors identify the reference point assumed by individuals with those individuals' *status quo* or current state.

Kahneman and Tversky go into great detail as to how reference points may vary, emphasizing that those reference points fixed by the *status quo* may shift as a result of encoding losses and gains relative to expectations that differ from the ones determined by the *status quo*. They also mention more specific cases where different encodings of the same pair of options create discrepancies between the reference point and the actual situation. According to them, this is exactly what happens when the choice is encoded in terms of final outcomes, as suggested from decision theory, instead of in terms of losses and gains (cf. Kahneman and Tversky 1979, 286–287).

A variation in the way a message is encoded, therefore, entails a change of context that has both cognitive and motivational consequences. Such consequences will depend on the kind of encoding that is being used. Considering all forms of framing effects detected so far would go beyond the scope of the present paper, which is limited to the so called "valence framing effects", that is, those effects resulting from a positive or a negative encoding of a message. Scientific research into these kinds of effects has led various authors to try to complete the list of variables involved in processing different encodings, thereby explaining the corresponding framing effects. In addition to loss aversion, endowment, preservation of the *status quo* and the tendency to ignore similarities —all of them trends acknowledged by Kahneman and Tversky in their studies on risky choice framing—, Levin and his collaborators point to the activation of positive associations in memory as the main mechanism responsible for framing effects (cf. Levin et al. 1998, 164–165). Positive stimuli generated by a frame would yield some associative responses that, in turn, would cause a clear increase in the level of approval that each individual assigns to the positively described option as opposed to that assigned to the negatively described one. It has even been demonstrated that the mere activation of positive associations with respect to one of the options presented for a given choice brings about substantial positive distortions of that option against the other one (cf. Russo et al. 1996, 103–107). In the experiment on distortion of alternatives carried out by Russo and his co-workers, positive descriptions of the owner of a restaurant or hotel remarkably influenced the more positive evaluation of the restaurant or hotel, despite the fact that such descriptions were logically independent of the attributes of the products offered. These experimental results reveal the same confirmation bias related to selective attention mechanisms as the one that has been observed in more general studies regarding the effect of expectations on judgment.

Turning now to the attempts at explaining goal framing effects, it is worth stressing the strong empirical support for the hypothesis of the negativity bias (cf.

Taylor 1991, 68–71). According to this hypothesis, individuals pay more attention to negative information than to equivalent positive information, showing themselves more influenced by the former than for the latter. From the decade of the 1990s, some of the explanations for the different framing effects have been partially unified, more specifically, loss aversion is understood as a subclass of the negativity bias, and the *status quo* bias is in turn regarded as a subclass of the loss aversion bias. In all these cases, the rejection caused by a loss is higher than the desire to obtain a gain of the same magnitude (cf. Levin et al. 1998, 177).

10.3.3   The Pursuit of Control over Framing Effects

If, especially during the 1990s, the detection, classification, and explanation of framing effects constituted a challenge only partially overcome despite the efforts made to that end, the challenge of controlling such effects has hardly been addressed. Yet, the identification of different bias that are activated according to the kind of frame in use sheds some light on the way individuals process information depending on how the latter is presented to them.[9] The obvious consequence seems to be that, if a certain form of encoding the message is avoided, the bias caused by such encoding can be avoided too, and, together with it, the introduction of certain variable that detracts from the validity of the survey. All forms of validity – statistical as well as internal, construct and external– could be improved by avoiding the encoding responsible for the bias. Nevertheless, even if researchers decided to proceed this way, the question would remain of what the most neutral possible frame would be, or, to put it differently, what frame would be the least amenable of producing a biased response from the individual decoding the message.

From a pragmatic standpoint, that is, from a view primarily committed to methodological effectiveness, one option would be to examine, among those empirical findings obtained by experimenters working on framing effects, those which point to variables that diminish or prevent such effects. It is important to notice that the above findings have been very scattered and hardly ever replicated, since they have been obtained through studies not directly oriented to determine this sorts of variables, but rather focused on the detection of framing effects. However, despite the more basic goal served by these experiments, in some cases they included additions that turned out enlightening for the purpose of controlling framing effects. In the case of risky choice framing, for example, it was demonstrated that when some question about the subject's reasons for a certain choice was added to the survey, then the framing effect was diminished or even eliminated. It is what R.

---

[9]References to "the bias" induced by frames can be very often found in research literature, although in a narrow sense referring to a difference between observable traits of the respondent and what she or he reports (cf. Groves and Singer 2004, 38–39). This narrow concept of bias is not the one applied in the present context of discussion.

P. Larrick, E. E. Smith and J. F. Yates call "the reflection effect" (1992, 199), which, according to their results, would make it possible to reverse framing effects by means of reflection on the issue presented within the frame. In a similar vein, Stephen M. Smith and Irwin P. Levin experimentally showed that individuals with a lower need for cognition were more affected by framing effects than those with a higher need for cognition, who in turn where almost immune to differences in framing (Smith and Levin 1996, 283).[10]

Experimental results suggest that factors other than the above also have a bearing on the scope of framing effects. Among these factors there are the domain of problems presented, the traits of experimental subjects, the magnitude or probability of potential outcomes, and the categories applied in verbalizing such outcomes (Levin et al. 1998, 153). For instance, subjects are more inclined to take risks related to health issues than related to finances. The other two cases referred above, however, could be covered by the general case where the amount of information handled by the subject is inversely proportional to the scope of the framing effects (Schoorman et al. 1994, 520). As already observed, the variations in such amount may be due to variations intrinsic to the frame, and basically dependent on how detailed the frame is, or to variations in the subjects, mainly related to their need for cognition or degree of competence on the kind of subject presented. With respect to the traits of experimental subjects, it has been found, for instance, that experts or students in a certain field tend to be less affected by framing effects when confronted with options evaluable from such field. Similarly, it has been verified that replacing expressions like "many" or "few" with numerical values lowers the intensity of framing effects. In the study by Schoorman et al. referred earlier, it has been experimentally established that the subject's degree of involvement or responsibility concerning a given issue can also eliminate the bias produced by the framing of the issue. Moreover, some recent empirical findings show that the framing bias is eliminated when the implicit frame is presented explicitly both verbally and visually (Gamliel and Kreiner 2013; Kreiner and Gamliel 2016), or when the addressee's attention is drawn to it (Kreiner and Gamliel 2018). All these procedures would help reestablish reliability and validity by increasing the consistency in the answers collected, improving the robustness of statistical correlations, eliminating confounds, ultimately allowing for a better empirical grounding of constructs and a higher generalizability of both results and procedures.[11]

---

[10]Within the field of psychology, the need for cognition constitutes a personality variable reflecting the individuals' disposition to perform cognitive tasks that require effort.

[11]As Jiménez-Buedo points out, two different senses of generalizability —and thus of external validity— are usually mixed in the literature; one refers to the degree in which an experimental finding can be considered 'representative' of conditions outside of the experiment, and another points to the extent to which such finding can be applicable to parallel situations (cf. Jiménez-Buedo 2011, 276). Without questioning the problematic implications of such ambiguity, it seems that both statistical and internal validity are preconditions for external validity, since, otherwise, there would be nothing to generalize in either sense.

The above considerations can also be applied to the case of the bias caused by the attribute frame, even if, as noted earlier, the sort of effect produced by this frame is the most homogeneous and clearly verified among ones caused by the valence frames. Despite the different domains of problems or the differences between subjects, the positive description of an item attribute, as opposed to its negative description, will almost always favor the more positive evaluation of both the attribute and the corresponding item. However, also in the case of attribute framing, a lower intensity of the bias has been experimentally determined when there is, on the subjects' side, a high degree of involvement as to the issue being described (Marteau 1989, 90–93; Millar and Millar 2000, 860–863). We find here again a phenomenon that suggests an inverse relationship between the intensity of the framing bias and the level of processing of information provided to the subject. This phenomenon might, therefore, support the hypothesis, backed up by the experimental work of Durairaj Maheswaran and Joan Meyers-Levy (1990, 365), according to which the more involved an experimental subject is in the issue described, the more detailed his or her processing of the information related to the issue. Moreover, several experimental studies have shown the occurrence of a closely related phenomenon, namely, that the evaluation of real items is less affected by framing bias than the evaluation of hypothetical items. Attribute framing effects are also diminished when subjects are asked to explain their answers or give reasons for them, and can be eliminated when attributes are represented both positively and negatively, being also represented both verbally and visually by using charts (Gamliel and Kreiner 2013).

Let us finally briefly consider some possible factors relevant in the control of goal framing effects. Like in the former cases, the degree of involvement in the topic presented, together with the tendency of the subjects to make a cognitive effort, are inversely related to the intensity of the framing effect.[12] Perhaps because of the greater structural complexity of goal framing, there are more variations in operationalizing this framing, which ultimately entails a less homogeneous evidence for goal framing than for attribute framing (Levin et al. 1998, 176). More specifically, such operationalization can be done either through simple negation (not obtaining profits) or through alternative terminology (losing the possibility of obtaining profits). Even if it seems obvious that linguistic variation may influence the strength of all sorts of valence framing effects, there are more potential linguistic variations in the case of goal framing, since the latter involves describing the consequences ascribed to some behavior as opposed to those ascribed to not performing such behavior. As Levin and his co-workers emphasize, in order to clarify when the responses of the subjects are dependent on semantic variations, it is necessary to develop an empirical study on language itself (1998, 174).[13]

---

[12]Numerous references to empirical studies that point to this issue can be found in Levin et al. (1998, 174).

[13]In his paper from 1992, Rolf Mayer provides some early clues to develop the kind of study suggested above. There he refers to some semantic aspects relevant in framing effects, such as

Here we find another instance of methodological development connected to newly recognized problems arising from methodological solutions to previous problems.

As for the need to focus on language, it is worth mentioning that there have been some attempts at explaining framing effects in general on the basis of the traditional semantic distinction between extension (what is designated by an expression) and intension (the way of determining extension). From the field of philosophy of economics, for example, Ivan Moscati has recently argued for understanding framing effects as doxastic effects caused by the intensional discrepancy between extensionally identical descriptions. Surveys employed by Tversky and Kahneman in their experiments included extensionally equivalent descriptions of outcomes and probabilities which, nevertheless, intensionally differed by virtue of the way uncertainty was presented, either in one stage games or in two stages games (Moscati 2012, 7). Moscati points to the problem of referential opacity in intensional contexts as that which would explain the apparent irrationality of subjects' tendency to prefer one option over the other:

> If we look at framing effects using the notions of intension and extension, they no longer appear to be manifestations of irrationality. Rather, they seem to be just other instances of the failure of the substitutability principle in referentially opaque contexts. Therefore, when looked at from the intension-extension viewpoint, the relevant problem shifts from the issue concerning the individuals' rationality, to the question of whether standard, set-theoretic economic models are able to capture the intensional difference between extensionally equal objects (Moscati 2012, 8).

According to this author, the apparent manifestations of irrationality would be the consequence of an apparent co-extensionality, mistakenly taken as real by those researchers who overlook the opaque nature of intensional contexts such as that of subjects' beliefs.

Although following a different strategy, Sacha Bourgeois-Gironde and Raphaël Giraud (2009, 385–387) also make use of the distinction between intension and extension to explain how framing effects come to happen. Both authors draw attention to the fact that, in economic methodology, the principle of invariance or extensionality goes beyond the logical principle establishing the co-extensionality between expressions whenever the latter are interchangeable *salva veritate* (i.e., whenever truth-value is preserved). In the context survey research, what needs to be guaranteed by means of co-extensional descriptions is not only truth-value preservation but also the preservation of whatever information proves relevant for making decisions. What needs to be specified, therefore, is the kind of information regarded as relevant for purposes of deciding among the options presented. Only after such information had been specified, could framing effects be ascertained as violations of the extensionality principle in the contexts of decision under study. Violating extensionality would then imply that irrelevant information determines the choices or judgments made by the subjects.

---

the clustered nature of meaning, the impact of thematic roles or the distinction between discursive background and discursive front.

Bart Geurts' 2013 article on framing offers another insightful discussion of the linguistic implications of framing effects. In his view, frames support counterfactual reasoning of the sort: if a state of affairs is positively or negatively described, then a different, respectively less or more advantageous state of affairs could have been the case. An important innovation of Geurts' approach is the explanation of framing effects, not only in terms of alternatives, but also in terms of what he calls "alignment". Expressions like 'too' or 'even' would depend on alternatives for conveying the speaker's intended message. For instance, 'even φ' would mean that φ is true and that φ's prior probability is low, relative to φ's alternatives (Geurts 2013, 7). Such alternatives are ordered in a scale and being "stronger" in the scale could be expressed with'>'. According to Geurts, implicatures depending on ordered alternatives support automatic inferences about the correlation (alignment) between prior probabilities and strength (Geurts 2013, 8). The definition of alignment states that, for any ψ, ψ' that are included among φ's alternatives, if $\psi > \psi$ then ψ » ψ' (where 'ψ » ψ'' means that ψ is more improbable than ψ'). The intuition behind this definition can be expressed by saying that "'more' on the quantity scale entails 'more' on the improbability scale" (Geurts 2013, 9). An important point emphasized by this author is that the Alignment assumption is optional (thus not part of the lexical meaning) and operates by default on the basis of world knowledge (Geurts 2013, 10). Our regular exposure to correlations between quantitative and qualitative scales, together with our tendency to establish connections and pursuing coherence, would explain the emergence of alignment assumptions (Geurts 2013, 11). Framing effects would also be a manifestation of this combined phenomenon, they being the result of establishing connections between different frames and different counterfactual alternatives. In applying the above analysis to framing, Geurts arrives at an evaluative understanding of framing effects and, therefore, adds 'it is good that [φ]'in order to uncover the underlying alignment assumptions (with '»' now meaning 'is better than'). Imagine that an airplane with 600 passengers crashed and we hear that 300 people survived or, alternatively, that 300 people died.[14] Our default alignment assumption would automatically yield the following interpretation for the positively frame description: 300 people survived » n people survived (such that $300 > n$). Obviously, this interpretation would be inconsistent with our usual understanding of the negatively framed description, that is to say, we would reject that 300 people died » n people died (such that $300 > n$). As Geurts concludes, far from being equivalent, both descriptions convey mutually inconsistent information about counterfactual states of affairs (2013, 12).

From a more pragmatic and pluralistic standpoint, Manski has explored the possibility of overcoming the flaws of economic survey research by following the same methods as those applied in cognitive psychology, which mainly rely on the determination of expectations. Thus, partly relying on a methodological tradition coming from empirical psychology, this author argues that research procedures should include questions about subjects' predictions concerning their own future

---

[14]I am here slightly modifying Geurts' example for the sake of simplicity.

behavior or self-reports on their own way of making decisions (Manski 2004, 1330–1331). An overall more robust treatment of subjects' patterns of decisions would be achieved by combining two different kinds of evidence, on the one hand, the observed conduct of subjects in making decisions, and, on the other, the self-reports made by respondents. As a result, there would be a wide plurality of procedures employed to determine not only subjects' expectations, but also their preferences, cognitive habits and intentions. Even if bringing with them a whole array of new difficulties, the use of self-reports would enable researchers to estimate several aspects involved in framing effects. In particular, it would make it possible to uncover several potential confounds and hence estimate the amount and kind of information actually processed by the subject, as well as the latter's need for cognition and underlying interests or preferences varying his or her attention mechanisms.

As a consequence of the above, construct validity could be highly improved, since the theoretical explanation of how subjects make decisions would become both theoretically and empirically more detailed by specifying how the postulated preferences are constrained by different psychological aspects involved in the interpretation of the options offered to them. These aspects could be empirically determined to some extent on the basis of survey results, self-reports and observed behavior. Improving construct validity would directly strengthen statistical and internal validity as well. In both cases, it would be possible to isolate the effect of different variables (expectations, understanding of options) that were previously operating as confounds and, for this reason, were obscuring the possible statistical correlation or causal link between the independent variable (preference) and the dependent variable (decision).

10.4   Conclusions

The present paper has called attention to the continuous process of feedback between empirical problems and procedural solutions taking place in scientific research. It has been argued that such dynamics entails a methodological refinement that naturally leads to a pluralistic methodological development. I have emphasized how this development goes hand in hand with the possibility of improving the validity of empirical research. The case of survey research in economics has served to illustrate this kind of dynamics. Here the attempt at overcoming the shortcomings of the theory of revealed preference, more precisely, the need to determine those expectations involved in decision making, leads to the use of a wide variety of survey procedures in economic methodology. Such use makes it possible to determine the effect of expectations on preferences and the subsequent effect of preferences on decisions, thereby improving statistical and internal validity of correlations between preferences (independent variable) and decisions (dependent variable). This way of controlling variables, however, requires researchers to face the methodological challenge of detecting and controlling framing effects.

The previous discussion includes an overview of the main ways to account for framing effects. Some of the most recurrent variables, like loss aversion, positive associations triggered by positive descriptions or selective attention drawn by negative information, point to well entrenched tendencies in most individuals. Other variables, like subjects' degree of involvement, cognitive effort, explicitation mechanisms, implicatures or situated linguistic understanding, more directly reveal the importance of cognitive, semantic and pragmatic factors. The improvement of statistical, internal validity depends on the successful empirical determination of the above factors, and, therefore, on their mitigation as possible confounds. Construct validity, more in particular, the validity of the postulated causes for individuals' decisions, also depends on the identification of the abovementioned factors. In particular, to account for framing effects in terms of mere mistakes in understanding is a wrong approach, given all the evidence on such effects examined earlier. Instead of keeping a theoretical construct that empirically under-represents the causal factors directly involved in decision making, it would be necessary to enrich such construct by establishing an aggregate of variables able to empirically represent the phenomenon under study. Since preferences and expectations are not the only variables that prove causally relevant in decision making, the explanation of the latter should include a reference to the rest of variables already mentioned. Finally, since the above kinds of validity are preconditions for external validity, the latter can only be obtained after the former has been accomplished. The broad range of intricate survey problems to be addressed in the future no doubt will require a healthy dose of methodological pluralism.

References

Angoff, W. H. (1988). Validity: An evolving concept. In H. Wainer & H. I. Braun (Eds.), *Test validity* (pp. 19–32). Hillsdale: Lawrence Erlbaum Associates.

Bourgeois-Gironde, S., & Giraud, R. (2009). Framing effects as violations of extensionality. *Theory and Decision, 67*(4), 385–404.

Caamaño-Alegre, M. (2013). Pragmatic norms in science: Making them explicit. *Synthese, 190*(15), 3227–3246.

Campbell, D. T. (1957). Factors relevant to the validity of experiments in social settings. *Psychological Bulletin, 54*, 297–312.

Campbell, D. T. (1986). Relabelling internal and external validity for applied social scientists. In W. M. K. Trochim (Ed.), *Advances in quasi-experimental design and analysis* (pp. 67–77). San Francisco: Jossey-Bass.

Carmines, E. G., & Zeller, R. A. (1979). *Reliability and validity assessment*. London: Sage.

Claveau, F. (2011). Evidential variety as a source of credibility for causal inference: Beyond sharp designs and structural models. *Journal of Economic Methodology, 18*(3), 233–253.

Cronbach, L. J., & Meehl, P. E. (1955). Construct validity in psychological tests. *Psychological Bulletin, 52*, 281–302.

Dellinger, A. B., & Leech, N. L. (2007). Toward a unified validation framework in mixed methods research. *Journal of Mixed Methods Research, 1*(4), 309–332.

Downward, P., & Mearman, A. (2007). Retroduction as mixed-methods triangulation in economic research: Reorienting economics into social science. *Cambridge Journal of Economics, 31*, 77–99.

Franklin, A. (2005). *No easy answers: Science and the pursuit of knowledge*. Pittsburgh: University of Pittsburgh Press.

Galison, P. (1997). *Image and logic: A material culture of microphysics*. Chicago: University of Chicago Press.

Gamliel, E., & Kreiner, H. (2013). Is a picture worth a thousand words? The interaction of visual display and attribute representation in attenuating framing bias. *Judgment and Decision Making, 8*(4), 482–491.

Geurts, B. (2013). Alternatives in framing and decision making. *Mind and Language, 28*(1), 1–19.

Grice, H. P. (1975). Logic and conversation. In P. Cole & J. L. Morgan (Eds.), *Syntax and semantics* (Speech acts) (Vol. 3, pp. 41–58). New York: Academic.

Groves, R. M., & Singer, E. (2004). Survey methodology. In J. S. House, F. T. Juster, R. L. Kahn, H. Schuman, & E. Singer (Eds.), *A telescope on society: Survey research and social science at the University of Michigan and Beyond* (pp. 21–64). Ann Arbor: University of Michigan Press.

Haack, S. (2006). Introduction. In S. Haack (Ed.), *Pragmatism, old and new. Selected writings* (pp. 15–69). New York: Prometheus Books.

Hacking, I. (1983). *Representing and intervening. Introductory topics in the philosophy of natural science*. Cambridge: Cambridge University Press.

Hausman, D. (2012). *Preference, value, choice, and welfare*. Cambridge: Cambridge University Press.

Jiménez-Buedo, M. (2011). Conceptual tools for assessing experiments: Some well-entrenched confusions regarding the internal/external validity distinction. *Journal of Economic Methodology, 18*(3), 271–282.

Jones, M. K. (2007). A Gricean analysis of understanding in economic experiments. *Journal of Economic Methodology, 14*(2), 167–185.

Kahneman, D., & Tversky, A. (1979). Prospect theory: An analysis of decision under risk. *Econometrica, 47*(2), 263–291.

Kreiner, H., & Gamliel, E. (2016). Looking at both sides of the coin: Mixed representation moderates attribute framing bias in written and auditory messages. *Applied Cognitive Psychology, 30*, 332–340.

Kreiner, H., & Gamliel, E. (2018). The role of attention in attribute framing. *Journal of Behavioral Decision Making, 31*, 392–401.

Kuhn, T. S. [1962](1970). *The structure of scientific revolutions*. Chicago: The University of Chicago Press.

Larrick, R. P., Smith, E. E., & Yates, J. F. (1992). *Reflecting on the reflection effect: Disrupting the effects of framing through thought*. Paper presented at the meetings of the Society for Judgment and Decision Making, St. Louis, November, MO.

Laudan, L. (1981). A problem-solving approach to scientific progress. In I. Hacking (Ed.), *Scientific revolutions*. Oxford: Oxford University Press.

Levin, I. P., Schneider, S. L., & Gaeth, G. J. (1998). All frames are not created equal: A typology and critical analysis of framing effects. *Organizational Behavior and Human Decision Process, 76*(2), 149–188.

Maheswaran, D., & Meyers-Levy, J. (1990). The influence of message framing and issue involvement. *Journal of Marketing Research, 27*(3), 361–367.

Manski, C. F. (2004). Measuring expectations. *Econometrica, 72*(5), 1329–1376.

Marteau, T. M. (1989). Framing of information: Its influence upon decisions of doctors and patients. *British Journal of Social Psychology, 28*(1), 89–94.

Mayer, R. (1992). To win and lose: Linguistic aspects of prospect theory. *Language and Cognitive Processes, 7*(1), 23–66.

Mayo, D. G. (1996). *Error and the growth of experimental knowledge*. Chicago: The University of Chicago Press.

Messick, S. (1989). Validity. In R. L. Linn (Ed.), *Educational measurement* (3rd ed., pp. 13–103). New York: American Council on Education/Macmillan Publishing Company.

Millar, M. G., & Millar, K. U. (2000). Promoting safe driving behaviors: The influence of message framing and issue involvement. *Journal of Applied Social Psychology, 30*(4), 853–866.

Moscati, I. (2012). Intension, extension, and the model of belief and knowledge in economics. *Erasmus Journal for Philosophy and Economics, 5*(2), 1–26.

Nardo, M. (2003). The quantification of qualitative survey data: A critical assessment. *Journal of Economic Surveys, 17*(5), 645–668.

Pelham, B. W., & Blanton, H. (2003). *Conducting research in psychology. Measuring the weight of smoke.* Belmont: Wadsworth/Thomson Learning.

Popper, K. R. ([1991]1999). All life is problem solving. Reprinted in: K. R. Popper. *All life is problem solving* (pp. 99–104). London: Routledge.

Rescher, N. (1977). *Methodological pragmatism. A Systems-theoretic approach to the theory of knowledge.* New York: New York University Press.

Russo, J. E., Medvec, V. H., & Meloy, M. G. (1996). The distortion of information during decisions. *Organizational Behavior and Human Decision Process, 66*(1), 102–110.

Schoorman, F. D., Mayer, R. C., Douglas, C. A., & Hetrick, C. T. (1994). Escalation of commitment and the framing effect: An empirical investigation. *Journal of Applied Social Psychology, 24*(6), 509–528.

Schwarz, N. (1996). *Cognition and communication: Judgmental biases, research methods, and the logic of conversation.* Hillsdale, NJ: Erlbaum.

Searle, J. (1978). Literal meaning. *Erkenntnis, 13,* 207–224.

Shadish, W. R., Cook, T. D., & Campbell, D. T. (2002). *Experimental and quasi-experimental designs for generalized causal inference.* Boston: Houghton Mifflin.

Sireci, S. G. (2009). Packing and unpacking sources of validity evidence: History repeats itself again. In R. W. Lissitz (Ed.), *The concept of validity: Revisions, new directions, and applications* (pp. 19–37). Charlotte: Information Age Publishing.

Smith, S. M., & Levin, I. (1996). Need for cognition and choice framing effects. *Journal of Behavioral Decision Making, 9*(4), 283–290.

Starr, M. A. (2014). Qualitative and mixed-methods research in economics: Surprising growth, promising future. *Journal of Economic Surveys, 28*(2), 238–264.

Steinle, F. (1997). Entering new fields: Exploratory uses of experimentation. *Philosophy of Science, 64*(supplement), s65–s74.

Suppes, P. (1998). Pragmatism in physics. In P. Weingartner, G. Schurz, & G. Dorn (Eds.), *The role of pragmatics in contemporary philosophy* (pp. 236–253). Vienna: Hölder-Pichler-Tempsky.

Taylor, S. E. (1991). Asymmetrical effects of positive and negative events: The mobilization-minimization hypothesis. *Psychological Bulletin, 110,* 67–85.

Tversky, A., & Kahneman, D. (1981). The framing of decisions and the psychology of choice. *Science, 211*(4481), 453–458.

Tversky, A., & Kahneman, D. (1991). Loss aversion in riskless choice: A reference-dependent model. *Quarterly Journal of Economics, 107,* 1039–1061.

Wiener, P. P. (1973–1974). Pragmatism. In P. P. Wiener (Ed.), *The dictionary of the history of ideas: Studies of selected pivotal ideas* (Vol. 3, pp. 551–570). New York: Charles Scribner's Sons.

Chapter 11
The "Economic Method" and Its Ethical Component: Pluralism, Objectivity and Values in Amartya Sen's Capability Approach

Alessandra Cenci

Abstract This paper addresses critical methodological issues raised by the reductionism-simplicity and scarce concern for fairness of orthodox value-free economics by focusing on Professor Amartya Sen's ethical review of the foundations of welfare economics.

What is previously assumed is that scientific research or science-related policy-making in several research areas within economics, or in the social sciences more generally, ought to be considered as "activities oriented to ends" – thus, unavoidably dependent on value judgements and, explicitly, ethical judgements.

With the view of accomplishing the persistent demands of ethical and method-ological pluralism, Sen's "ethics and economics" paradigm, his human development model – the "capability approach" – and, especially, the adoption of the social choice theory as an overarching framework for social evaluation represent strong methodological challenges not only to standard utility-based economics, but also to mainstream liberal-egalitarian theories of justice. Indeed, Sen's strategy of introducing ethical evaluation – thus, a concern for a plurality of objects of practical value and for multiple ends of action – substantially enriches conventional economic accounts and powerfully enhances fairness-equity.

The major insights of Sen's "economic method", entailing a strong ethical component, are expected in the applied research of areas where the shortcomings of reductionist-monist economics have been mostly detrimental, such as social evaluations of wellbeing or health and global justice.

Keywords Reductionism · Pluralism · Complexity · Objectivity · Fairness-equity · Economic evaluation · Global justice

A. Cenci (✉)
University of Southern Denmark (SDU), Odense, Denmark
e-mail: alessandra@sdu.dk

© Springer Nature Switzerland AG 2020
W. J. Gonzalez (ed.), *Methodological Prospects for Scientific Research*, Synthese Library 430, https://doi.org/10.1007/978-3-030-52500-2_11

11.1 Introduction

One of the central discussions in the twentieth-century philosophy of science is about the tension among views that consider science and scientific inquiry as value-free or, conversely, as value-ladenness activities. The value-free ideal (VFI) of science – which entails the acceptance of the neo-positivist separation between facts and values, as well as high degrees of ethical and methodological reductionism – has been the dominant research paradigm for a long time within both the social sciences (SS)[1] and in economics.

The underlying view is that value-freedom and/or ethical-methodological reductionism should have guaranteed the unity of science, and that reductionist tenets should simplify the tasks of theorisation and operationalisation by increasing the manageability of the complexity of the real social world.

In economics, the acceptance of formal, highly idealised models based on a set of unreal assumptions regarding human behaviour, rational agents, contexts of choice and so on has been functional to accomplish these goals. Since Robbins ([1945] 2007), the economic science is explicitly concerned with the VFI and the fact-value distinction. There is total separation between ethics and economics, and the supposed relationship between the two disciplines is of mere juxtaposition (Hausman 1992). This standpoint is paradigmatically represented by the differences between the methodology of positive and normative economics Friedman ([1966] 2007). Positive economics is value-free and its capacity to predict (i.e., advance scientific knowledge) is substantially attached to the formal axioms of the standard rational choice theory[2] (RCT) and to a normative concept of rationality entirely grounded on epistemic values, such as *simplicity*, *logicality*, *clarity*, *universality*, and *consistency* (Gomez Rodriguez 2009). This is the philosophical basis of orthodox economics[3] and the related "economic method", which has been the dominant interpretation also in the field of normative economics (Hausman and McPherson 2007, 226–250). Moreover, the "economic approach" through the RCT became one of the prominent paradigms for social explanation in several disciplinary fields such as political science, sociology, cognitive psychology etc.[4]

Nevertheless, the last few decades seem to have been characterised by a fundamental paradigm shift. There is an effort to move away from the VFI of science and associated reductionist positions – in fact, efforts to embrace scientific

---

[1] A still influent view of objectivity in the social sciences is the one defended by Max Weber ([1904] 2007).

[2] For an overview of standard RCT, see (Reiss 2013, Ch. 3, 29–53).

[3] The terms *orthodox* economics and *neoclassical* economics can be used interchangeably since refer to the same approach to values and normative judgements in economics.

[4] An overview on how the reductionist economic approach (functionally based on the standard RCT) has been adopted in several disciplinary areas, see (Mari-Klose 2000).

pluralism[5] are increasing, both in the natural sciences and in the SS (Kellert et al. 2006). As evidenced by Gonzalez (2013, 1505), this renewed stance relies on the idea that science and scientific research are "*human activities oriented to ends*" and, thus, unavoidably context-dependent and deeply influenced by the values and socio-cultural norms of specific scientific or policy settings. In this scenario, another fundamental discussion is whether and how scientists make use of *value judgements* when doing science, and what is the nature and scope of value judgements in scientific inquiry.[6] More recently, both the VFI and the unique interest in the epistemic aspects of science and scientific inquiry (i.e., epistemic values)[7] have been questioned (Kincaid et al. 2007). Accordingly, the direct or indirect role that non-epistemic aspects (i.e., non-epistemic values) have or should have in scientific research and in knowledge production in democratic settings has become a fundamental matter (Longino 1990; Kitcher 2001; Douglas 2009). As opposed to traditional views, both epistemic and non-epistemic values are thought to be *intrinsic* and *foundational* to scientific inquiry and science-related policymaking. These developments are also present in economics, where the VFI and the fact-value dichotomy seem no longer defensible (Putnam 2002; Putnam and Walsh 2012; Reiss 2017). A crucial issue still at debate is whether and how similar value-sensitive science and knowledge could maintain acceptable standards of *objectivity* – i.e., to what extent can values of ethical and social importance be incorporated by scientific models without affecting the epistemic status of science and the validity and reliability of scientific knowledge attained.

What this paper substantially argues is that the overall work of Professor Amartya Sen in the fields of welfare economics and decision theory is relevant to advance ongoing debates about reductionism and pluralism in scientific inquiry. In particular, Sen's re-interpretation of orthodox economics and economic analysis can be considered significant examples of *ethical* and *methodological pluralism*.

This stance openly contrasts the *ethical reductionism*; namely, the monism inherent to standard utility-based economics which has largely been used to justify controversial tenets such as the *identifiability* of utility and well-being or the total *commensurability* of every aspect valuable for well-being in utility terms. This ideal has been fundamental to validate a reductionist "economic method" that pays exclusive attention to utilitarian information and individual utility-based welfare achievements.

---

[5]*Scientific Pluralism* is the philosophical view that aims to demonstrate the viability that some phenomena require multiple accounts. Pluralists observe that scientists present various, sometimes even incompatible, models of the world and argue that this is due to the complexity of the world and representational limitations that, in any case, need to be addressed and not circumvented by adopting scarcely empirically useful reductionist and highly simplified exploratory, explanatory, predictive models.

[6]For an overview of the on-going debate concerning values, science and democracy in the so-called *Politics of Science* see, (Machamer and Wolters 2004).

[7]On the primacy of epistemic value see Laudan (1984), McMullin (1982), Dorato (2004).

Sen openly rejects not only the VFI but also *monism* to further embrace *informational complexity* and *value pluralism*, as well as adopting the methodologies of different disciplines – namely, economics, normative ethics, political theory, and empirical SS – to consider aspects of the social reality previously neglected by economic accounts (Sen 2009b). The resulting view is paradigmatically expressed in *on ethics and economics* (Sen 1987) and by his human development model, the capability approach (Sen 1999a). They represent, in fact, pivotal contributions to the development of an "ethical economics" and a rigorous "economic method" that is sensitive to a plurality of (non-epistemic) values of ethical and social importance. Therefore, Sen's *system of social evaluation* might go beyond reductionism-simplicity-monism and scarce concern for the fairness-equity of orthodox utility-based economics. But most importantly, it might successfully cope with acknowledged practical-operational flaws of standard economic analysis and measures in the view of obtaining more accurate and reliable scientific knowledge.

These claims are illustrated by focusing on two critical cases – namely, research areas in which the shortcomings of reductionist-monist utility-based economics and its related methodology have been considered mostly detrimental, such as in social evaluations of wellbeing or health[8] and redistributive justice, especially at a global level. The paper aims to draw attention on the ability of Sen's "economic method", incorporating a strong ethical component, to re-address critical issues that are hard to solve by adopting conventional frameworks for social evaluation (i.e., welfare economics) or global justice (i.e., liberal-egalitarian theories). The emphasis is on the epistemic and normative import and, hence, on the far-reaching practical implications (i.e., methods applied) of Sen's review of the foundations of neoclassical economics and decision theory, the view frequently called *liberal welfarism* for its exclusive attention to individuals' welfare achievements.[9] What is believed is that Sen's ethical revision of standard RCT's notions of rationality, the rejection of value-free/value-neutral interpretations of objectivity as well as the amendment of the formal conditions of rational social choice (respectively Sects. 11.3.1, 11.3.2, 11.4) have substantially re-shaped orthodox economics' individualistic tenets and methodology, both in a pluralist and contextual sense.

While Sen's account of rationality has extensively been discussed in the literature, other aspects of Sen's thinking – particularly, the importance of the adoption of the *social choice theory* (SCT) as an overarching framework for social evaluation and distributive justice – are still largely unexplored (Comin 2018, 181). This paper aims to fill this gap by uncovering the close interconnection between Sen's capability approach (Sen 1999a), the capability justice (Sen 2009a) and

---

[8]Standard economics analysis and methodology – either in welfare or health evaluation – have been seen to rely on the same epistemic and normative tenets (Anand et al. 2004; Hausman 2015). Thus, in this chapter, when referring to evaluation of welfare and wellbeing also allude to health evaluation.

[9]*Welfarism* is the view underlying orthodox economics. Here, wellbeing, health is usually measured by means of survey-based methods eliciting individuals' preferences. Preferences are then, merely sum-aggregated at collective level (i.e., by applying sum-ranking).

their social choice foundations. The emphasis is on the insights for creating a "social" perspective functionally based on genuine social preferences (not mere sum-aggregations of the individual's ones as in standard economics accounts). In contrast to traditional reductionist and individualistic interpretations of science and economics, Sen's "ethical" method have incorporated value and judgements pluralism to encourage meaningful informational expansions at the applied level. As a result, recent empirical research in the target fields (i.e., multidimensional appraisals of welfare-health-justice) has notably advanced by explicitly adopting Sen's *extra-welfarist* tenets[10] (i.e., ethical rationality, multidimensional welfare, a mix of deontological and consequential evaluation). A key challenge for capability-based multidimensional assessments is evading reductionism and scarce concern for fairness-equity of standard economics measures while maintaining acceptable scientific standards. Thus, another fundamental task of this paper (conclusive section) is to clarify how earlier supposed dilemmas with the epistemic status of economics as a science (i.e., rationality, objectivity-impartiality of beliefs, claims, value judgements behind operational choices or evaluative practices), normally attached to the inclusion of *non-epistemic values* and *value judgments pluralism*, has been handled by latest developed capability-based accounts and measures. Specifically, how rigour, validity and reliability of scientific inquiry and scientific knowledge might be preserved when adopting normatively richer evaluative-comparative frameworks such as Sen's capability approach.

The paper develops as follows:

Section two (11.2) links fundamental methodological problems in welfare economics with the adoption of a value-free/value-neutral notion of rationality (i.e., standard RCT) and restrictive formal conditions of rational social choice (i.e., Arrow's SCT). In particular, standard RCT's narrow assumptions regarding agents, contexts, and motivations for action underlying rational decisions, both at an individual and a collective level, have often been considered mainly responsible for the scarce exploratory and explanatory capacity or predictive potential of orthodox economic theory, laws, and models. Nonetheless, standard RCT's notion of rationality has been largely used at an applied level to validate *reductionism-simplicity-monism* of utility-based economic accounts (i.e., one-dimensional measures of wellbeing and health, evaluative procedures exclusively based on economic-monetary variables, exclusive attention to utilitarian information and consequential evaluation). Thus, the adoption of Sen's method, including a strong ethical component, is supposed to profitably move away from the main practical-operational flaws of standard economic approaches – namely, reductionism and scarce concern for fairness-equity. Besides, it is extremely important for redistributive justice and public policymaking in democratic settings (e.g., allocations of "scarce" public resource, priority setting). Indeed, the same reductionist behavioural model (i.e., RCT) is espoused also by mainstream liberal-egalitarian theories of justice and

---

[10]For an overview of the main methodological differences between *welfarist* and *extra-welfarist* accounts of wellbeing and health, see (Brower et al. 2008).

determines their incapacity to properly address concrete fairness-equity claims, especially at the global level (also in Sect. 11.4).

Section three (11.3.1, 11.3.2) focuses on the insights of Sen's economics and method' reviewed foundations to build a *virtue-ethics approach* functionally based on plural values of ethical and social importance and incorporating *ethical evaluation* into standard economic analysis. On the one hand, the section illustrates how Sen's ethical revision of the behavioural foundations of orthodox economic theory – namely, "ethical" rationality (1977) have been used to expand the *informative basis* of welfare economics and thus, improve practical applications in the field. The capability approach and correlated multidimensional measures of wellbeing-health-justice are the main results of this theoretical expansion (Sect. 11.3.1). On the other hand, the section clarifies that a similar broader idea of rationality cannot properly function without also rejecting value-free/neutral interpretations of objectivity and further assuming a notion that is sensitive to agents' plural values and contexts of choice (Sect. 11.3.2). This espousal is supposed to expand the epistemic and normative basis of *objective value judgements* and incorporate *agents' diversity* and *value pluralism* while protecting the integrity of scientific inquiry (in economics and in the SS) and scientific knowledge production from subjectivism or moral relativism.

Section four (11.4) elucidates why SCT has been chosen by Sen as the ideal overarching framework for multidimensional, pluralistic social evaluations (obtained by applying the capability approach) under conditions of diversity and value heterogeneity typical of contemporary complex societies. What is believed is that Sen's (and others') review of the formal conditions of rational social choice validates "objective-impartial" value judgements leading to "plural but objective values". The resulting view is largely in keeping with the notion of objectivity exposed in Sect. 11.3.2 and what ultimately corroborates the capability approach's extra-welfarist tenets and their intrinsic pluralism. Indeed, by accommodating value and goals pluralism and their *non-commensurability*, a reviewed SCT can be more respectful of people's diversity, their capacity for self-determination and agency (Sen 2009a, ch. 4; Comin 2018, ch. 8).

Section five (11.5) brings together the final remarks about the behavioural, epistemic and non-epistemic foundations of the overall methodology underlying Sen's system of social evaluation and his proposal for global justice. The section concludes by referring to some latest applications of Sen's concepts in economic evaluations of public health in which, the key dilemmas attached to normative approaches and ethical evaluation, are significantly minimised by adopting specific evaluative tools. What is finally suggested is that *normatively richer* measures of wellbeing or health and social justice can highly be consistent with the production of *objective scientific knowledge.* Thus, conceivably, that a more systematic implementation of similar complex, value-sensitive and pluralistic economic models and evaluative methods, could directly contribute to improve societal outcomes.

11.2 Reductionism Versus Pluralism in Economics and Social Choice Theory

Insistent demands for ethical and methodological pluralism – against value-free, the legitimate determinants of rational behavior (Sen 2005) reductionist economics and the related "economic method" – have pervaded the economic thinking since its origins. In fact, the idea of a total separation between ethics and economics not only contrasts with foremost positions maintained within classical philosophy, which did not conceive of economics as separated from politics or morality (e.g., Aristotle),[11] but is also rather inconsistent with key arguments advanced by eminent economists such as Adam Smith in the *Theory of Moral Sentiments* (Sen 1987, ch. 1).

A (frustrated) attempt to modify the relationship between economics as a science and ethical values is found in Kenneth Arrow's pioneering reflection on the formal criteria of SCT.[12] He incorporates some explicit and externally imposed ethical desiderata within the formal requirements of rational social choice. What he obtains is the so-called "impossibility theorem" that establishes the inconsistency of a few ethical constraints with recognised formal conditions of rationality (i.e., completeness, transitivity).[13] Even though the formal result obtained by Arrow's research was not new,[14] it has been largely (and often improperly) understood, at least among economists, as the impossibility of reconciling the demands of rationality with those of morality. It has been interpreted as the undeniable proof of the irreconcilability and total separation of ethics and economics, as well as of the unfeasibility of making interpersonal comparisons among people's levels of satisfaction (Arrow 1963, 101–103).[15] In other words, Arrow's theorem reiterates a standard interpretation of economics as an "economic science", in which there is wide distance between technical-scientific and ethical matters (e.g., Robbins's). Accordingly, values and particularly ethical values have "legitimately" been expunged by economic theory, laws, and models through the adoption of a value-free notion of rationality, the one of the standard RCT. Standard RCT's value-free assumptions (i.e., its axioms) are intended to guarantee the "scientific" character of the discipline since, by allowing the "economic method" to focus on factual

---

[11]This view is expressed by Aristotle mostly in the "*Politics*" and in the *Nicomachean Ethics*.

[12]Social choice theory (SCT) is a theoretical framework for analysis that combining individual opinions, preferences, interests, or welfares reaches *collective decisions* or *social welfare* evaluations (e.g., regarding wellbeing or justice). It has been criticised for using elements of formal logic (axiomatic method) and especially, for being methodologically individualistic

[13]This event is central to the development of Sen's research program. Sen began his academic activity by attempting to solve Arrow's paradox (with scarce success) by weakening the formal requirements of rational social choice with a special focus on *transitivity* (see Sen 1969).

[14]Arrow's theorem shows the same formal result of the Condorcet's "voting paradox" dated 1785. An overview of the general debate in Public Choice in (Mueller 2003).

[15]Arrow's model is based on *ordinal utility measurements* those prevent to assess the *magnitude-intensity* of people's levels of satisfaction, thus, formally established the impossibility of making interpersonal comparisons of people's welfare formerly argued by Lionel Robbins.

judgements only, increase the possibility to obtain objective-impartial scientific knowledge. This vision is precisely, what justified the adoption of reductionist tenets and methodology.

However, as evidenced by Gonzalez (1999), the omission of values is especially harmful in the field of *applied science*. Thus, it is not surprising that some of the main practical-operational flaws in *welfare economics* and *economic evaluation* has been related to standard RCT's notion of rationality and the narrowness of the underlying behaviour model – i.e., the homo economicus (Sen and Williams 1982; Sen 1982; Rescher 1988; Green and Shapiro 1994; etc.). Particularly, lack of *realism*[16] of RCT's assumptions but also *monism*, have been indicated as to be especially detrimental for accuracy, validity, and reliability of explanations, predictions, and related prescriptions (Sen 1986; Sent 2006). Indeed, the absence of (ethical) values and the acceptance of self-interested utility maximisation as the unique rational end of action determinate conventional economic analysis' exclusive attention for *utility measurements* and *consequential evaluation* (Sen 1979; Hansson 2007; Hausman 2015). That is, by supposing the *total commensurability* of every aspect relevant for individuals' wellbeing in utility-monetary terms; in *one-dimensional* reductionist analysis (i.e., cost-benefit, cost-effectiveness, cost-utility), the utilitarian information (i.e., utility rankings) is regularly the only knowledge informing scientific research or public policymaking (e.g., resource allocation-distribution, priority setting).

In this scenario, Amartya Sen's seminal work in welfare economics and SCT represents a strong effort in improving the exploratory and explanatory capacity and predictive potential of economic theory and economic models. Briefly, by incorporating ethical motivations (i.e., ethical values) into standard RCT's notion of rationality, Sen aims at expanding orthodox economics models, methods and related measures of wellbeing and health or social justice. The result is a broader evaluative-comparative space (the capability approach) that, by being further inserted into an all-encompassing decision-making framework (the SCT), is expected to meaningfully embody *complexity* as well as *value* and *judgements pluralism* with no issues for rigorous and objective-impartial knowledge production. Extra-welfarist analytical concepts like the *multidimensional welfare* (Sen 1985a) symbolize the concern for the multiple aspects of human wellbeing and the plurality of human ends for action and directly challenge reductionist utility-based economic accounts and measures exclusively based on consequential evaluation. Sen's proposal is a comprehensive system of social evaluation – namely, a *goal-rights system* functionally based on both deontological and consequential evaluation and reasoning in which reflexions about procedural aspects such as agent's diversity, real social contexts, or the extent of positive freedom that people actually enjoy in alternative states of affairs are also explicitly involved (Sen 2000b; Gotoh 2014). Similarly, Sen's deontological-consequential economic method represents a huge challenge to the contemporary contract theory tradition in the field of redistributive

---

[16]On realism and "scientific realism" in economics, see Lawson (1997) and Mäki (2011).

justice. Sen's strategies differ substantially from liberal-egalitarian theories of justice's emphasis on ideal theorising, hypothetical contracting and their exclusive reliance on deontological evaluation when establishing principles and normative ideals of redistributive justice underlying just societies and institutions. In addition, mainstream liberal-egalitarian theories of justice (Rawls 1970; Nozick 1974) accept standard RCT's reductionist assumptions regarding human behaviour. Thus, their challenge to liberal welfarism (i.e., utility-based theories) seems insufficient to provide alternative frameworks to improve allocative and redistributive practices both at the local and the global level.

In more detail, here is Sen's research agenda developed throughout different but interconnected steps:

First, the rejection of the homo economicus and the "ethical" revision of RCT's notion of rationality (Sen 1977) is the strategy adopted by Sen to introduce "theoretically" *complexity* and *value pluralism* – namely, a concern for a plurality of objects of practical values – into standard evaluative practices of many branches within economics. These reflections gave origin to the Capability approach that, by incorporating *non-utilitarian information* and *ethical-deontological evaluation* into standard cost-benefit analyses (Sen 2000a), expands the *informational basis* of welfare economics and current measures of wellbeing-health, quality of life, poverty and so on.

Second, there is the formulation of a notion of objectivity (i.e., "positional" objectivity) that can be sensitive to agents, contexts and plural values but still satisfactorily based on objective-impartial value judgements. Sen's notion openly clashes with traditional value-free/value-neutral concepts of objectivity, and denounces their inaptness to validate real choices, especially those of an ethical-normative nature, under conditions of social and ideological pluralism in specific social, cultural, and policy environments. Major insights of Sen's concept rely on the supposed nature and scope of objectivity-impartiality and objective-impartial value judgements underlying social evaluations or global justice claims. In fact, Sen's concept establishes that *scientific objectivity* cannot properly be achieved by abstracting from the contingencies of the subjects expressing value judgements or by ignoring the ineliminable plurality of values and norms of concrete social-policy environments.[17]

Third, the revision of the formal conditions of rational social choice (SCT) undertaken by Sen (and others) in later years is fundamental to validating reviewed notions of rationality, objectivity and correlated methodological enhancements in social evaluation or global justice appraisals. As opposed to traditional accounts (i.e., Arrow's SCT), the possibility of a *rational social choice* under conditions of human diversity, social and ideological pluralism is established by assuming the feasibility of intersubjective rational deliberation on values and a practical reason that most profitably deals with epistemic or moral disagreement.

---

[17]This statement is consistent with several positions defended in *social epistemology* see Longino (2019).

The next sections address the three points separately by evidencing the main differences between Sen's ethical approach, his method and orthodox utility-based economics. Likewise, Sen's methodological differences with mainstream liberal-egalitarianism – primarily, John Rawls' *theory of justice as fairness* (1970) and its subsequent evolution at a global level (1999)– are also highlighted.

11.3 The Foundations of Sen's Economics

11.3.1 Ethical Rationality

As a point of departure, Sen argues that although scientific inquiry in economics and in the SS more generally inevitably requires simplifications, it ought to be possible to be rigorous without being reductionist. This possibility is seen to have a close relation with, and is highly reliant on, the adoption of specific conceptions of rationality and rational behaviour.

According to Sen (1985b, 341): "The choice of behavioural assumptions in economics tends to pull us in two different-sometimes contrary-directions. The demands of tractability can conflict with those of veracity, and we can have a hard choice between simplicity and relevance. We want a canonical form that is uncomplicated enough to be easily usable in theoretical and empirical analysis. But we also want an assumption structure that is not fundamentally at odds with the real world, nor one that makes simplicity take the form of naivety. There is a genuine conflict here-a conflict that cannot be easily disposed of either by asserting the need for simplification in theorizing or by pointing to the need for realism. What we have to face is the need for discriminating judgment, separating out the complications that can be avoided without much loss and the complexities that must be taken on board for our analysis to be at all useful."

As renowned, several practical-operational flaws in economic evaluation and social research have been related to the formal axioms of standard RCT's notion of rationality, particularly to its normative character and the lack of realism in its underlying assumptions. However, moving against proposals for the revision of standard rationality that are aimed at enhancing the descriptive character of the theory (by increasing the realism of its assumptions),[18] Sen not only advocates for maintaining the normative character of standard rationality but further promotes enlarging its moral basis. In his view, a descriptive standpoint, exclusively based on instrumental reasons, is incapable of describing the width of agents' positive freedom (their "agency") or the inescapable plurality of human ends. Proper

---

[18]Proposals for increasing the realism of RCT's assumptions entail either considering "real" rational agents' cognitive limitations as in Simon's *Bounded Rationality* or, the influence of the context of choice when taking rational decisions as in Kahneman's and Tversky's *Prospect Theory*. An overview of the overall debate in Sen (2009a, ch. 8).

accounts of rationality and rational behaviour ought to evaluate not only the means, but also the ends of action-choice since not all human ends (and underlying values) can be considered equally or automatically valid. In other words, what seems to be needed is a rationality of ends that, by entailing a broader rational scrutiny, could provide better assessments of people's wellbeing, their quality of life, or the extent of positive freedom that they enjoy in alternative states of affairs. Sen's interpretation of rationality seems to converge with those of other eminent scholars – for instance, with Nicholas Rescher's understanding of a *rationality of ends* profitably contrasting the "economic rationality" (Rescher 1988, ch. 6). Essentially, Rescher imagines a normative notion that, once applied, *"can avoid adaptive preferences"* – which is the main problem of utilitarian evaluations – since it is *"structurally open to the inclusion of plural reasons to act"* (99–105).

Thus, according to Sen (2002, 4): "rationality should be interpreted broadly, as the discipline of subjecting one's choices – of actions as well as of objectives, values priorities – to reasoned scrutiny. Rather than defining rationality in terms of some formulaic conditions that have been proposed in the literature (such as satisfying some specified axioms of "internal consistency of choice" or being in conformity with "intelligent pursuit of self-interest" or being some variant of maximizing behaviour), rationality is seen here in much more general terms as the need to subject one's choices to the demands of reason." Particularly, collective decisions are seen as deeply influenced by many social factors such as the membership of a particular social group, property, the interest regarding the community.

What Sen criticises are three main understandings of the notion of instrumental rationality in economics (Sen 2002, 19): (a) rationality as internal consistency of choice, (b) rationality as self-interested behaviour, and (c) rationality as maximisation in general. By taking positive economics as an example, Sen clarifies that making predictions – based on a set of unreal assumptions about contexts, agents' behaviour and so on – is not the only task within economics. Indeed, the discipline intrinsically possesses a prescriptive character, and prescriptive analysis demands keeping in mind an idea of "the good and the bad" (Sen 1986, 3). In other words, it demands identifying "What should it be?" or answering the fundamental Aristotelian question of "How should we live?". However, accomplishing these tasks imply going beyond the VFI of economics; hence, abandoning both the narrow RCT notion of rationality and the concept of self-interested behaviour (i.e., the maximisation of individual utility as the uniquely "rational" end of action-choice). Value-free models could hardly be representative of real choices in specific contexts in which ethical motivations or social norms also play a pivotal role.[19] Therefore, providing better descriptions, predictions, and prescriptions fairly demands explicitly considering ethical values, the plurality of human ends and how freely they can be pursued by rational agents in real settings. This interpretation openly challenges the reductionism-simplicity-monism of standard economics, as

---

[19]For examples of rational choices constrained by ethical and social norms see, (Sen 1993a).

well as emphasises the insights of complexity and plurality to obtain improved assessments.

According to Sen (1986, 22): "supplementing the postulates of self-interested behaviour by other (more cooperative) behaviour patterns complicates economic modelling but this may nevertheless facilitate the understanding of economic phenomena and also facilitate our ability to predict. Simplicity is not a virtue if it happens to be dead wrong. (. . .). The axioms of purely self-interested behaviour make economic modelling simpler but rather useless in many contexts."

Therefore, in Sen's view, economics as a science and scientific inquiry in the discipline entail evading the reductionism of the "economic" method by rejecting the formal axioms of RCT and revising economic rationality towards a more ethical interpretation (Sen 1977, 1994). An "ethical rationality" discards self-interest as the unique and rational course of action-choice and goes beyond self-interested objectives by including *sympathy*[20] and especially *commitment* among the legitimate determinants of rational behaviour. Establishing the logical-practical possibility of commitment allows us to reject orthodox economics' exclusive focus on competitive behaviour and recognise cooperation or solidarity as main products of rationality (and not as mere rationality failures). Likewise, an ethical rationality corroborates the existence of *positive duties* towards others who become moral subjects with the same rights and are no longer mere parameters (as in standard RCT).

What became also noticeable is that descriptions, explanations, and predictions based on individual preferences[21] over a set of given options and exclusive attention towards utility rankings could be insufficient to provide accurate assessments of agents' *positive freedom* and their welfare and social justice achievements in specific contexts (Sen 1987, 1–22). It is precisely the introduction of *ethical motivations* into standard RCT rationality and of *ethical evaluation* into standard economic analysis that allows Sen to overcome the reductionism-simplicity-monism of standard economics with the view of building a genuine *virtue-ethics approach* (Van Stavaren 2007; Burbidge 2016). Sen's strategy of enriching the behavioural foundations of economic theory in an ethical sense is what generates an ethically sensitive but rigorous "economic method" that could substantially enhance the exploratory and explanatory capacity and predictive potential of standard economic models and measures (Sen 1987, 110). Recent methodological trends in welfare economics and economic analysis such as *multidimensional-multicriteria evaluation* and *multidimensional indicators research* explicitly rely on Sen's concepts and

---

[20]Sen considers that, differently from commitment, sympathy is in keeping with standard economics behavioral assumptions since consistent with the self-interested behavior. His interpretation comes from Adam Smith (Sen 1987, ch.1).

[21]For an overview of the on-going debate on individual preferences in economics, see Reiss 2013, Ch.12.

method to validate broader assessments based on a set of analytical variables of qualitatively different nature.[22]

All in all, an explicit consideration of the plurality of agents' values and rational ends of action is what delivers an expanded evaluative-comparative framework for social evaluation and, thus, more complex and accurate assessments of welfare and health or social justice. Demonstrably, taking rational decisions at a societal level – for example, allocating or redistributing "scarce" resources both efficiently and fairly within a just society – entails examining the correctness, legitimacy and/or social desirability of a plurality of conflicting values and rational ends in specific social-cultural-policy settings. This is precisely what can be attained by adopting *multidimensional* analytical-comparative frameworks such as the *capability approach* (Sen 1999a), which is functionally based on a concept of *multidimensional welfare*, either in social evaluations of wellbeing and health or in assessments of social justice both at a local and a global level (Sen 2009a). The capability approach's broader normative tenets and its intrinsic attention to procedural considerations regarding agents, contexts shift attention away from utilitarian information and allows economic analyses to consider multiple variables of different natures, and also of an ethical nature (Sen 1982). The *informational diversity* and underlying value pluralism clearly distinguish Sen's capability approach from orthodox economics' monist and utility-based theories and related measures, as well as from liberal-egalitarian theories and related accounts of justice. In a capability-based assessment, wellbeing, equality, and social justice are measured according to what people can actually do or be – namely, according to their capability-freedom intended as actual opportunities to achieve "the things that they have reasons to value" (Sen 1999a, 32–33). Decisive attention is given to the multiple determinants of people's wellbeing and to self-chosen realisations that, as such, might enhance agents' positive freedom, their capacity for self-determination and agency. For all these reasons, Sen's capability approach is often indicated as realising a fundamental *paradigm shift* in welfare economics – namely, from welfarism to extra-welfarism (thus, beyond "monist" utility-based assessments and metrics). Likewise, it is also known to offer a renewed perspective on the "currency" of distributive justice that enriches existing liberal notions of "equality" (see Kaufman 2005).[23] Therefore, capability-oriented evaluations based on a set of self-chosen and qualitatively different valuable objects, might represent an *alternative* standpoint to either any kind of subjective utility-based account (e.g., happiness,

---

[22]Popular capability-based multidimensional measures are the inequality adjusted *Human Development Index* (HDI) and, the *Multidimensional Poverty Index* (MPI) both implemented by UNDP (United Nation Development Program).

[23]According to Sen (1980), a truly egalitarian society ought to provide *"equality of capabilities"* and not, for instance, equality of welfare or equality of resources.

desire, preferences) or expert-led so-called objective-list theories,[24] in which too little space is left for public debate and democratic deliberation (Sen 2004).

To conclude this theoretical appraisal, a foremost philosophical challenge when defending Sen's *procedural-deliberative* approach – functionally based on a wide-ranging, pluralistic-based, ethical rationality – against standard economics and mainstream liberal-egalitarian thinking is to demonstrate whether and how multiple objects of practical values and/or dimensions for evaluation-comparison could meaningfully be selected by rational agents in democratic settings. A crucial task relies on the identification of a *rational social choice procedure* suitable for validating a plurality of satisfactorily objective-impartial values and goals. In contemporary complex societies, a similar choice procedure is undoubtedly challenged by the presence of conflicting "good" values and rational ends for action. Thus, validating *value pluralism* and *objective value judgements* at a collective level, most likely, demands a notion of objectivity that support the possibility of conflicting "objectively good" reasons.

A similar idea of objectivity is presented in the following sub-section and is central to the justification of Sen's "ethical" account and its underlying value pluralism and value judgement pluralism.

11.3.2 *"Plural-Contextual" Objectivity and Rational Deliberation on Values*

Although Sen's arguments on rationality convincingly demonstrated (also to ortho-dox economists) that *ethical values* pervade even the core of economic theory; within the "economic science", the reflections on objectivity (i.e., objective beliefs, claims, judgments, knowledge) have been rather impermeable to these evolutions. However, the soundness of a "rationality of ends", and, of Sen's "ethical rationality" relies on the actual possibility of having *objective values*. In traditional views of objectivity, making *objective value judgements* have been largely attached to the possibility of providing de-contextualised and/or impersonal reasons, such as in Thomas Nagel's *The View from Nowhere* (1986) or Nicolas Rescher's formulation of objectivity (1997).

Sen's approach to objectivity and objective-impartial judgments goes in a completely different direction. His aim is to demonstrate, instead, the far-reaching implications of what he calls *parametric dependence* or *positional objectivity* for collective action and rational social choice under conditions of diversity and plurality, for scientific debates, decision theory, ethics and public affairs.

---

[24]Popular objective-list theories are Rawls' theory of justice as fairness (i.e., the five *Primary Goods*) and Martha Nussbaum's version of the *Capability Approach* (2000). They are briefly discussed in Sects. 11.4 and 11.5.

According to Sen (1993b, 126): "what we can observe depends on our position vis-à-vis the object of observation. What we decide to believe is influenced by what we observe. How we decide to act relates to our beliefs. Positionally-dependent observations, beliefs and actions are central to our knowledge and practical reasons. The nature of objectivity in epistemology, decision theory and ethics has to take adequate note of the parametric dependence of observation and inference on the position of the observer."

That is, people's notion of good, their beliefs are unavoidably dependent on their intrinsic characteristics (living context, personal experience, own condition, vested interests, etc.). So far, if two people are in the same position and share the same values, their evaluations and judgements should be identical. Their evaluations or judgements are *positionally objective*.

What Sen clarifies is that this *positionally dependent* view of objectivity is rather important not only in normative choices, but also in scientific debates. Indeed, positionally objective claims and judgements represent an inevitable preliminary phase of resultant *trans-positionally objective* scientific interpretations and beliefs (Sen 1993b, 130–131). Differing from standard views or other recently formulated "contextual" notions of objectivity,[25] Sen's concept of *trans-positional objectivity* substantially relies on a *synthesis of*, and not an abstraction from, the peculiarities of the subjects, their beliefs and partial observations. The influence of the context is expressed by previously identified positionally objective judgements that are supposed to be functionally conducive to scientific beliefs and objective scientific knowledge. Another fundamental difference with standard accounts of objectivity is that objective scientific judgements do not solely rely on facts or empirical evidence but also, on a plurality of non-epistemic values of ethical and social importance. That is, the *fact-value dichotomy* is profitably reconciled[26] since values are established as legitimate objects of rational scrutiny and deliberation. Therefore, "plural contextual values"[27] can *intrinsically* and *internally* be incorporated (via ethical rationality) into economic models and functional measures without being detrimental for the objective-impartial judgements underlying scientific inquiry and scientific knowledge production thus, the knowledge informing science-related policymaking.

As palpable, major dilemmas for a similar *agent-context-dependent* and *value-sensitive* notion of objectivity come from decision-making situations that involve huge socio-ideological pluralism and value heterogeneity. The two critical cases indicated in the beginning of this chapter (i.e., welfare-health evaluation and global justice) are representative of the difficulties entailed in taking rational decisions

---

[25] For an opposite view of acontextual objectivity still based on abstraction from the peculiarities of subjects expressing value judgement and, suggest focusing on fact or evidence instead (e.g., Write 2018).

[26] For an analogous position see Putnam 2002, Part II.

[27] On the possibility of "contextual" values as elements informing "objective" knowledge production see, Cenci (2019).

in contemporary complex societies or in the global arena (i.e., public decisions involving conflicting individuals' values and vested interests).

Against traditional positivist views, Sen defends the actual possibility of the rational deliberation of values, as well as the idea that rational decisions under conditions of complexity, diversity and plurality could be based on "plural but objective values" chosen intersubjectively (e.g., by supposing a collective and, perhaps, transcultural practical reason). In his understanding, the choice of a set of "objectively good" values or ends of action demand to evaluate the *general defensibility* of the "goodness" of their underlying reasons. Essentially, for values and goals to be valid, they must survive an open, well-informed, intersubjectively conducted rational scrutiny (Sen 2009a, 385–387). Specifically, "values underlying capabilities should be the product of interpersonal moral activities and their acceptability relies on the underlying freedoms they represent" (Sen 2009a, 399). Analogous deliberative-validating procedures have been defended by Hilary Putnam and Thomas Scanlon (Sen 2009a, 30–35). For instance, Scanlon (1998) argues that the soundness of either moral beliefs or ethical values depends on the validity of the reasons provided in intersubjective reasoned deliberations, since objective values represent what the "*other agents cannot reasonably reject*". According to similar interpretations, a plurality of conflicting values can *rationally* and *objectively-impartially* be selected. Even though *value disagreement* can never totally be circumvented, value judgements underlying normative decisions could maintain acceptable degrees of objectivity. Thus, the resulting views can be invulnerable to the moral relativist objection as well as to accusations of subjectivism. The main strength of Sen's idea of objectivity relies, precisely, on it being based on a deeper contextual understanding of objective beliefs, values, goals thus, it can profitably reassess subjectivism, cultural relativism, as well to investigate gender or cultural bias in specific social circumstances.

Undeniably, Sen substantially diverges from conventional interpretations concerning the role of values – explicitly, non-epistemic values – either in normative reasoning or in scientific inquiry (in economics and in the social sciences). Traditionally, ethical values and value pluralism have been considered detrimental for rational behaviour, objective science and scientific knowledge production while moral disagreement is widely regarded, both in decision theory and ethical theory, as a failure of the practical rationality. In all these fields, a valid rational deliberation entails to achieve *complete agreements* on ethical principles and foundational values in all circumstances, often by postulating the possibility of a *unanimous consensus*. This is how Rawls (1970, 1999) supports his theory of justice as fairness, also at a transnational level: by assuming the unavoidability of a reflective equilibrium (RE) and an overlapping consensus regarding principles of justice and their underlying values, under the (fictive) conditions of objectivity-impartiality created in the original position (OP) by devices such as the veil of ignorance. These are the fundamentals of *Political Liberalism* (1993), which is widely accepted in contemporary ethical-political theory (even by Martha Nussbaum's version of the capability approach) and to which Sen's procedural-deliberative approach and the related ideas of objectivity and objective value judgements represent a crucial methodological advancement.[28] As pointed out by Anderson (2003) and Peter (2012), Sen's notions (i.e., positional and trans-positional objectivity), are valuable not only enlarge the foundations of welfare economics and related evaluative methods, but also to build versatile, normatively richer frameworks suitable to substantiate transnational theories of global justice and democracy.

However, a robust validation for Sen's account and evaluative method– based on value pluralism, informational diversity and "contextualised" objectively good reasons established by their critical defensibility in a rational interpersonal scrutiny – demands rational choice foundations that diverge substantially not only from the ones of mainstream liberal thinking but also from those supposed by orthodox economics and decision theory (i.e., RCT, Social choice theory). The next section concentrates on Sen's (and others') review of the formal conditions of *rational social choice* and stresses its potential to perform agent-context-value-sensitive social evaluations of welfare and health or, global justice in contemporary complex societies.

11.4 Sen's Social Choice Theory (SCT): Validating Informational Diversity, Value Judgement Pluralism, and Moral Disagreement

This section elucidates the close relation between a reviewed SCT and the epistemological, normative, and operational solutions defended by Sen over the years (i.e., ethical rationality, plural-contextual objectivity, multidimensional welfare, informational diversity, inclusion of non-utilitarian variables into standard economic analysis). It deals with how Sen's overall proposal to abandon the reductionism-simplicity-monism of orthodox utility-based economics and related troubles in practical applications allows to generate a comprehensive analytical and normative space for social evaluations of welfare-health and redistributive justice, especially at a global level. Special emphasis is placed on how reviewed formal conditions of rational social choice would represent a fruitful stratagem to better handle fairness-equity claims under conditions of human diversity, and social and value pluralism – i.e., a heterogeneity of beliefs and value judgements that is typical of contemporary complex democracies and globalised societies.

Following Sen (2009a, intro), developing new frameworks of reasoning with regard to social evaluation or global justice entails addressing the same methodological problem: superior social evaluations and social justice appraisals are closely dependent on the type of information included or excluded from value judgements. As shown in the former sections, the informative basis adopted in practical

---

[28]For a critical assessment of Sen's deliberative method against Rawls' RE, and how this is in keeping with Sen's interpretation of a *Goals-Rights System* functionally based on both *deontological* and *consequential* evaluation and reasoning see Gotoh (2014).

applications within economics or in the social sciences has severely been constrained by the reductionist, epistemological and normative assumptions of orthodox economics. Then, reductionist tenets directly impacted on evaluations-comparisons and measurements obtained since these are exclusively based on the knowledge extrapolated by the individual's utility rankings, merely aggregated at collective level. Thus, Sen's concepts of rationality or objectivity – both incorporating value and goals pluralism and contextual-procedural considerations – demonstrate the far-reaching implications of the broadening of the informational basis of welfare economics or, redistributive justice. It is the joint adoption of ethical-deontological and economic-consequential evaluation as well as the inclusion of non-utilitarian variables that has expanded conventional cost-benefit analyses and provided more complex, multidimensional and accurate accounts of wellbeing or justice – namely, according to the extent of positive freedom-capability, agency actually enjoyed by diverse people in alternative states of affairs.

However, what is still "philosophically" challenging about this supposed "pluralist" perspective is demonstrating how rational decisions under conditions of diversity, value judgments pluralism and epistemic or moral disagreement could actually be taken at a societal level, since, in standard SCT, the logical-practical possibility of doing so is denied by the *formal conditions of rationality* extensively accepted by a long tradition that can be traced back to the enlightenment (Condorcet, Borda, and more recently, Arrow). Differently from mainstream decision theory and ethical theory, Sen endorses that even if value dilemmas or moral conflicts could not be totally eliminated, these problems cannot be solved "methodologically" either by adopting reductionist tenets and methods (orthodox economics) or, by postulating a practical reason capable of improbable unanimous consensus and complete agreements on foundational values (liberal theories of justice). The conclusion is that a genuine pluralist, non-reductionist account – i.e., underlying the capability approach and a capabilitarian framework for redistributive justice – necessitates reviewed formal conditions of *rational social choice* in which human and informational diversity, as well as social and value pluralism (hence, *noncommensurability* of different values) should most profitably be accommodated (Osmani 2009, 2010).

First of all, establishing the consistency of a plurality of values of ethical and social importance with rational behaviour as well as the plausibility of multiple rational ends of action by rejecting the standard criteria of rationality and rational choice are vital preconditions. In line with Sen's ethical review of the standard RCT's notion of rationality, what orthodox economists or other social scientists (Arrow 1963; Becker 1976 etc.) indicate as key characteristics of rational behaviour – such as the consistency of choice (i.e., the formation of a stable set of preferences over given options or self-interested utility maximisation as the only rational end of action-choice) – cannot be the exclusive criteria for rational choice irrespective of any change in the external circumstances (Sen 1993a, 2002). Demonstrably, ethical considerations and social norms are "external points of view" that act from the outside by constraining choices in many deliberative situations in concrete contexts. Indeed, when an individual understands that his choice limits other valuable possibilities or that his choice goes against recognised social codes, he might "reasonably" change not only his own immediate choice, but also the underlying criteria entailed by it (see examples of Sen 2002, 65–118, Sen 2009b, 1–35). Common phenomena such as akrasia, strategic behaviour, changes of preferences over time, and moral conflicts, which are represented in standard economics as mere rationality failures, demonstrate that it is problematic to constrain real behaviours within the narrow structure of both standard RCT and SCT (i.e., their formal axioms and reductionist assumptions underlying notions of the *instrumental rationality*) see, Sen (2009a, ch. 8).

According to Sen (1999b, 349), some fundamental questions in SCT refer to: how can it be possible to arrive at cogent aggregative judgements about the society (for example, about "social welfare" or the "public interest", or "aggregate poverty"), given the diversity of preferences, concerns, and predicaments of the different individuals within the society? How can we find any rational basis for making such "aggregative judgements" as the society prefers this to that, or the society should choose this over that, or this is "socially right"? Is reasonable social choice at all possible, especially since, there may be "as many preferences as there are people"?

Properly answering these questions entails reflecting on the formal conditions of *rational social choice*. Particularly, Sen identifies in the emphasis on *ordering completeness* – both in standard decision theory and ethical theory – what actually impedes meaningfully considering human diversity, value pluralism and value disagreement, as well as their impact on real choices in concrete social settings. Conversely, he sees in the admission of *partial agreements* and *incomplete orderings of preferences* what could make a social choice based on genuine "social preferences" possible (Sen 1982, 74–83 and 203–222). These ideas are consistent with the rejection of the *internal consistency of choice*[29] but especially rely on the acceptance of *incompleteness* as a legitimate formal requirement for rational social choice. In Sen's understanding (Sen 2018), incompleteness is important since it can properly represent a structural lack of information-knowledge (i.e., epistemic inadequacy), agents' diversities and the plurality of their values and rational goals for action. In other words, incompleteness is what properly represent the *incommensurable* structural differences existing among diverse people, their own values and preferences.

The insights of the reviewed formal conditions of SCT, implying the possibility of rational deliberation on values and the plausibility of *incomplete agreements* on a plurality of values of ethical and social importance, are particularly obvious when comparing Sen's approach to decision-makings applying Rawls' RE.[30] There are "hard" social choices in which value conflicts arise from the simple fact

---

[29] It implies the rejection of the formal condition of *transitivity* (if A is better than B, B is better than C, thus, A is better than C) that denies the possibility of *preference changes over time, strategic behaviour* or *counter-preferential choice*.

[30] Decision making procedures selecting principles of justice that relies on RE are still based on the *standard rationality*. That is, in the OP under a veil of ignorance, self-interested rational agents choose the principles of justice by following prudential rather than ethical motivations. Thus, the

that alternative options are based on normative views grounded in conflicting – but equally valid – values or principles of justice.[31] This is actually the case in prudential decisions and ethical judgements underlying a theory of justice. In all these situations, value conflicts inevitably arise but could be unresolvable under a choice procedure defending the idea that a unique and complete agreement on foundational values or principles of justice (a RE!) ought to be achieved in all circumstances. With special regard to moral reasoning and normative choices, what this standpoint disregards is that there could be a variety of plausible answers and, thus, the possibility of reaching a unique and complete agreement is rather unfeasible (Sen 2018, 12–13). In Sen's view, the solution is a "weaker" practical reason that does not postulate the necessity of reaching an agreement and, as such, ensures that a plurality of diverse values or normative ideals could *simultaneously coexist* in the same deliberative situation and thus, stay available to be actually chosen. Even though accepting incompleteness means reducing the scope of the practical reason, it does not imply a decisional or valuational impasse (e.g., as in Arrow's model). Rather, a SCT entailing a *less demanding* practical reason is expected to better endorse values and judgements pluralism. At an applied level, it ratifies the concern for *incommensurable variables* of qualitatively different natures enhancing fairness-equity. Therefore, accepting incompleteness is what allows to refute the reductionism-simplicity-monism of standard economic methodology and validate value-based complex, plural, and multidimensional accounts and measures.

The difference between a "maximum" and an "optimum" option – i.e., the difference between an *acceptable* and a *perfect* choice – is an essential element in Sen's argument for incompleteness and the request for a less demanding theory of practical reason. Sen's interpretation openly contrasts with widespread beliefs – both in decision theory and ethical theory – that reasoning about values and choice procedures remains unfinished until an optimal alternative has been identified (Sen 2017). Rather, he defends that a successful enough closure of a reasoning process identifies a maximal alternative that is not judged to be worse than any other available option – i.e., a *"maximal element"* that does not need to be optimal in the sense of being *"the best alternative"* but should be just "as good as every other available option". In set theory,[32] this is the definition of a partial ordering (i.e., meta-rankings based on incomplete preferences).[33] Once translated into Sen's vocabulary, a partial ordering is the mathematical representation of the

---

adoption of Rawls' theory perpetuates most of the epistemic, methodological, normative problems described in this paper at Sects. 11.2 and 11.3.

[31] See the cases described by Sen of *three children and a flute* (2009a, 12–15) or of *Ashraf's choice* (2018, 12–13).

[32] For a formal argument and justification for maximal choices by adopting the *set theory* see, (Sen 1997).

[33] For a discussion on Sen's notion of "meta-ranking" and how it better deals with categories such as *comparability, partial comparability, full comparability*, see Comin (2018, 186–191). For a formal justification of the *partial comparability* underlying Sen's notion of meta-ranking, see Munda (2016).

valued "capability set" – namely, a comparative-evaluative space yielding at least a maximal element that is available to agents to be chosen.

An example, frequently used by Sen (2002, 17; 2018, 15), might better clarify the meaning and scope of his proposal. It relates about the Buridan's ass who died of starvation dithering between two haystacks unable to decide which one was better. Likely, the ass could not have founded "the best" option since the haystacks could not be ranked vis-á-vis each other; but still, he had the opportunity of doing much better than starving to death. Either haystack would have been a maximal choice, and choosing a maximal alternative, even though not the best, would have been sensible enough.

What this example shows is that without additional *externally imposed* evaluative-comparative criteria (i.e., value judgements, normative elements), one alternative cannot be judged "better" or "worse" than another since it is impossible to establish the exact magnitude of "goodness" or "badness" embodied by the available options. To properly assess the available alternatives, decision-makers need a *normative ideal*, an *ethical desideratum* that is distinguished and distinguishable from the formal features of the comparison itself and, as such, could function as a guide in the evaluative-comparative exercises. In this vein, what the example additionally shows is that formal conditions of rationality, such as the consistency of choice, self-maximisation and completeness, can be *unnecessarily restrictive*. In fact, the ass could have chosen any one of the two alternatives in front of him – both were better than starving to death. Due to the fundamental lack of information (incompleteness), the ass could not have chosen the best option but, given the circumstances, every non-optimal (maximal) choice would have allowed him to achieve better welfare outcomes than the non-decision option that sentenced him to death. Therefore, what Sen's overall reflection on SCT's foundations has shown is that the search for the best option is often unable to properly measure the *width of a person's opportunities*, her *welfare options* or the *extent of positive freedom* and *agency* that people enjoy in alternative states of affairs. Most of the real-life situations are non-optimal, while real choices are infrequently based on solely discriminating among optimal alternatives. Thus, although the acceptance of incompleteness entails giving up the possibility of making optimal choices, it does not represent a big problem for SCT, since it opens up the path to a larger set of analytical-practical possibilities and increases the *realism* and *utility* of investigations.

On the whole, Sen's weakening of the formal conditions of rational social choice not only validates expanded frameworks for social evaluation, but also encourages the re-assessment of leading principles and regulative ideals for redistributive justice (both in economics and mainstream liberal thinking). On the one hand, controversial redistributive criteria in orthodox economics such as the *Pareto optimality* can be abandoned. Pareto's criterion by preventing the comparison of non-optimal alternatives impede to carry out interpersonal comparisons of wellbeing, thus, also the transfer of resources in favour of the worst off (Sen 1970, Ch. 2; Hausman and McPherson 2007). On the other hand, theories of *perfect justice*, due to biased redundant and unrealistic standards of completeness, are proved to be flawed in their practical uselessness.[34] The methodological differences between Rawls' and Sen's frameworks for global justice – what Sen called, respectively, a *transcendental approach*[35] and a *comparative approach* (Sen 2009a, 126–141) – exemplify the advantages of grounding social justice appraisals on a reviewed SCT. A capabilitarian approach to global justice and human rights[36] shifts attention away from unattainable perfect justice ideals, while the *patent injustices* arising from real social contexts could be addressed and conceivably, solved (Sen 2009a, Ch. 2). Here, (non-optimal) alternative societal arrangements are compared and ranked from "less just" to "more just" in order to get direct guidance on what to do. In line with previously defined notions of rationality and objectivity, Sen defends a comprehensive evaluative-comparative approach to justice (2012a, b), in which contextual considerations and a plurality of conflicting values, normative principles and alternative goals can *rationally* and *objectively-impartially* be scrutinised and then selected by adopting democratic deliberative methods. This is what increases the focus on fairness-equity and, plausibly, is able to improve societal outcomes in concrete socio-cultural-policy settings by means of adjusted public policies.

11.5 Final Remarks

In this paper, we have extensively illustrated the pillars of Sen's ethical economics and of a *capability-based system of social evaluation* – functionally grounded in the ethics and economics paradigm and in a reviewed SCT. We also explained how a similar broader analytical framework for social evaluations of welfare, health or redistributive justice -thanks to its improved foundations – is expected to solve the main methodological flaws of reductionist-monist utility-based economics but also of mainstream liberal-egalitarian theories of justice.

Differently from conventional accounts, Sen's capability approach allows deeper critical examinations into whether people in specific contexts and under certain circumstances could have the freedom to function in ways that matter to themselves (agency freedom) and their wellbeing (wellbeing freedom).[37] The resulting epistemic-normative view – functionally erected upon the ethical rationality, positional and trans-positional objectivity, a reviewed SCT – entails explicit concern for the contexts of choice and for a plurality of (non-epistemic) values of ethical and social importance while, at an applied level, demands the adoption of the method-

---

[34]For an assessment of Sen's critique of Rawls' *Transcendental Institutionalism* as based on a *redundancy* and an *unfeasibility* argument, see Osmani (2010).

[35]As widely acknowledged, the design of *perfectly just institutions* in a well-ordered society" is Rawls' main regulative ideal (a deontological "optimum") and the basis of his procedural justice approach.

[36]On the contribution of Sen's approach to human rights research, see Vizard (2005).

[37]For examples, see Sen (1985).

ologies of different disciplinary areas. Therefore, Sen's *procedural-deliberative system of social evaluation* and Sen's *comparative realisation approach to justice*[38] can be considered highly representative of interpretations in which the recurrent demands for ethical and methodological pluralism pervading the philosophical and scientific arenas are fulfilled. Undeniably, Sen's procedural-deliberative version of the capability approach entails vital epistemic and non-epistemic aspects such as *complexity, multi-dimensionality, vagueness, non-determination* that mirror the increased concern for *diversity, value pluralism*, as well as for people's *positive freedom* and *agency*. The combination of these elements generates a *flexible, versatile*, and *open framework* for social valuation and redistributive justice capable of re-addressing contextualised problems without supposing unnecessary "artificial" constraints (Alkire et al. 2008). In particular, the defence of an *informationally pluralist approach* that does not entail a clear indication of what objects of practical value free rational agents would designate in an open and objective-impartial social choice procedure is what distinguish it substantially from conventional approaches and that is valuable for several reasons.

First, "contextualised" civil societies are supposed to have the right-duty of selecting their own living principles and values through public debate and democratic deliberative procedures. This stance contrasts with both subjective preferences-based utilitarian accounts (welfare economics) and expert-led objective lists (political-ethical theory), which are either based on unrealistic assumptions or on predefined normative principles and regulative ideals. For example, Nussbaum's version of the capability approach – which is grounded on an *over-specified* list of 10 valuable capabilities for human life – does not substantially diverge from Rawls' interpretation and is vulnerable to the same objections: expert-led, over-specified, too rigid, western-oriented, based on a demanding and unrealistic deliberative method.[39]

Second, the established possibility of a *rational social choice* – validating a set of non-epistemic values of ethical and social importance that are rationally-objectively selected under conditions of human and informational diversity and socio-ideological pluralism – is a major contribution of Sen's reviewed SCT not only to the foundations of welfare economics, but also to ongoing debates on values, science and democracy. Resulting social evaluations or redistributive justice appraisals can be based on a plurality of "objectively good" values – meaningfully selected in an objective-impartial and rigorous manner at a societal level – in concrete social, scientific or policy settings. Here, the legitimacy of selected values

---

[38]These descriptions have been introduced by Osmani (2009, 2010).

[39]By focusing on *meta-ethical foundations*, I have argued elsewhere against the suitability of leading approaches in ethical theory (and in favor of Sen's deliberative approach) to substitute standard economics utility-based theories and measures for public policy (see Cenci 2011, 2015). In particular, both Rawls' and Nussbaum's' account accept the ethical-political view called *Political Liberalism* (Rawls 1993) that apply *reflective equilibrium* as decision method when selecting the valuable objects of the list or the principles of justice in a "well-ordered" liberal society.

and normative ideals underlying science and scientific practices in democratic settings largely depend on the epistemological and normative solutions outlined in Sects. 11.3 and 11.4. They represent the improved rationale of an *ethical economics* systematically based on the free exercise of democracy and public reasoning in which the rigour, validity and reliability of *scientific inquiry* and *knowledge production* necessitate the participation of the entire civil society. Thus, a strong precondition is that rational social choices can occur under *transparency* and by having the adequate information to take *free* and *well-informed* decisions.[40]

Some major dilemmas for Sen's complex and pluralistic approach might persist at the practical-operational level. Traditionally, a main source of justification for standard economics' reductionist assumptions– particularly monism – thus, for one-dimensional, utility-based measures of wellbeing-health-justice can be found in their ability to simplify the tasks of *operationalisation.* Tenets such as the identification of wellbeing and utility, the commensurability of every aspect valuable for individual or collective wellbeing in utility-monetary terms have been used to facilitate *comparison* and *aggregation* of qualitatively different analytical variables or, to justify their *quantification.* Conversely, the complexity, plurality and the deliberative tenets underlying Sen's version of the capability approach are notoriously challenging for building "workable" measures (Alkire et al. 2008; Coast et al. 2015; Karimi et al. 2016). As anticipated, multidimensional assessments of wellbeing, health, inequalities, and so on – openly inspired by Sen's research into the foundations of welfare economics – are, obviously, valuable in relation to overcoming the reductionism-simplicity-monism of standard economics' utility-based measures and to enhancing fairness-equity. As evidenced by Munda (2016), multidimensional accounts, by increasing complexity and enclosing value pluralism, intensify the accurateness and realism of explanations, descriptions, predictions supplied to policymakers. As further suggested by Arneson (2016), multidimensional measures – based on a plurality of objects of practical value – better support fairness-equity claims in subsequent evaluations and policymaking. Residual worries refer to the possibility of meaningful aggregating and comparing qualitatively different dimensions and an increased subjectivism (perhaps, arbitrariness) of valuations based on multiple objects of values that can be "weighted" rather differently by analysts or policymakers in analogous studies.[41] Due to quite a few possible interpretations and, the absence of a standard procedure to assign "weights" to qualitatively different evaluative dimensions, multidimensional assessments are known to frequently deliver inconsistent results. It explains why latest capability-

---

[40]For example, *experts's advisory boards* but especially, *scientific communication* become fundamental to take rational decisions on complex topics of general interest at societal level.

[41]A popular methodology to handle (i.e., to aggregate and compare) multidimensional measures is the so-called *weighting schemes* approach which is also common in popular Capability-based metrics (Alkire and Foster 2011). By still supposing the *total commensurability,* analysists assign a *numerical value* to qualitatively different dimensions in order to obtain a single, manageable measurement of wellbeing, health, poverty etc. for a given population and thus, build the rankings accordingly.

oriented research in the field of economic evaluation concentrates on how the *objectivity-impartiality* of judgements behind operational choices and evaluative practices could be increased to guarantee the *cogency* of experimental results. In this task, (Cenci and Hussain 2020) show how genuine *multidimensionality* and *non-commensurability* – namely, complexity and plurality – could profitably be embodied by functional measures of wellbeing and health without the usual problems associated the achievement of objective science and knowledge. This study illustrates how the objectivity-impartiality of value judgments behind operational choices made under the broader capability approach's extra-welfarist tenets can be raised substantially. The solution applied to circumvent typical practical-operational problems of extra-welfarist, non-consequentialist, multidimensional assessments is the adoption of tools such as the so-called *robust model evaluation* and *robust methods* that do not require subjective or arbitrary value schemes and, can simultaneously be based on plural "contextual values" and objective-impartial judgements.[42] Major benefits are expected for more precise identifications of the worst off and their objective needs in specific social-policy settings that could be addressed by focused public policies.

To conclude, Sen's economic method, entailing a strong ethical component, can be regarded as a crucial contribution to overcoming the reductionism-simplicity-monism of orthodox utility-based economics and corroborates ethical and methodological pluralism in *applied* economic and social research. Analogously, the *pluralistic normative* grounding of the capability approach (also established by adopting a reviewed SCT) is able to deliver workable measures of wellbeing, health, justice that can fruitfully be used to reassess and validate *unbiased science* and *objective scientific knowledge production* in contemporary complex democracies and in the global arena.

Bibliography

Alkire, S., & Foster, J. (2011). Counting and multidimensional poverty measurement. *Journal of Public Economics, 95*(7–8), 476–487.

Alkire, S., Comin, F., & Qizilbash, M. (2008). *The capability approach. Concept, measures and application.* New York: Oxford University Press.

Anand, S., Peter, F., & Sen, A. (Eds.). (2004). *Public health, ethics and equity.* Oxford: Oxford University Press.

Anderson, E. (2003). *Sen, ethics, and democracy.* Retrieved: http://www.personal.umich.edu/~eandersn/SenEthicsDemocracy.pdf. Accessed 3 Oct 2019.

Arneson, R. (2016). Does fairness require a multidimensional approach? In M. Adler & M. Fleurbaey (Eds.), *Oxford handbook of well-being and public policy* (pp. 588–614). Oxford: Oxford University Press.

Arrow, J. K. (1963). *Social choice and individual values.* New York: John Wiley and Sons Press.

---

[42]On the contribution of "contextual values" to objective knowledge production, see Cenci A. (March 2019).

Becker, G. (1976). *The economic approach to human behaviour*. Chicago: University of Chicago Press.

Brower, W. B. F., Culyer, A. J., Van Exel, N. J. A., & Rutten, F. F. H. (2008). Welfarism and extra-welfarism. *Journal of Health Economics, 27*, 325–338.

Burbidge, D. (2016). Space for virtue in the economics of Kenneth J. Arrow, Amartya Sen and Elinor Ostrom. *Journal of Economic Methodology, 23*(4), 396–412.

Cenci, A. (2011). Economía, Ética y libertad en el enfoque de las capacidades. *Revista Laguna, 29*, 123–147.

Cenci, A. (2015). Measuring well-being for public policy: A freedom-based approach. *Ágora, 34*, 111–144.

Cenci, A. (2019). "Contextual Values" and objective knowledge: Insights from applied economics. In: Paper presented at the *Danske Filsofisk årsmode*, March 2019.

Cenci, A., & Hussain, M. A. (2020). Epistemic and non-epistemic values in economic evaluations of public health. *Journal of Economic Methodology, 27*(1), 66–88. https://doi.org/10.1080/1350178X.2019.1646922.

Coast, J., Kinghorn, P., & Mitchell, P. (2015). The development of capability measures in health economics: Opportunities, challenges and progress. *Patient, 8*, 119–126.

Comin, F. (2018). Sens capability approach, social choice theory and the use of rankings. In F. Comin, S. Fennell, & P. B. Anand (Eds.), *New frontiers of the capability approach* (pp. 179–197). Cambridge: Cambridge University Press.

Dorato, M. (2004). Epistemic and nonepistemic values in science. In P. K. Machamer & G. Wolters (Eds.), *Science, values and objectivity* (pp. 52–77). Pittsburgh: Pittsburgh University Press.

Douglas, H. (2009). *Science, policy, and the value free ideal*. Pittsburgh: University of Pittsburgh Press.

Friedman, M. ([1966] 2007). The methodology of positive economics. Reprinted in: D. M. Hausman (Ed.), *The philosophy of economics: An anthology* (pp. 145–178). Cambridge: Cambridge University Press.

Gomez Rodriguez, A. (2009). The micro-foundations of social explanation. *Epistemologia, 32*, 5–22.

Gonzalez, W. (1999). Ciencia y valores éticos: De la posibilidad de la Ética de la Ciencia al problema de la valoración ética de la Ciencia Básica. *Arbor CLXII, 638*, 199–171.

Gonzalez, W. (2013). Value ladenness and the value-free ideal in scientific research. In C. Luetge (Ed.), *Handbook of the philosophical foundations of business ethics* (pp. 1503–1521). Dordrecht: Springer.

Gotoh, R. (2014). The equality of the differences: Sen's critique of Rawls's theory of justice and its implications for welfare economics. *History of Economic Ideas*, xxii/2o14/1.

Green, D., & Shapiro, I. (1994). *Pathologies of rational choice theory: A critique of applications in political science*. New Haven: Yale University Press.

Hansson, O. S. (2007). Philosophical problems in cost-benefit analysis. *Economics and Philosophy, 23*, 163–183.

Hausman, D. (1992). The inexact and separate science of economics. In *Cambridge*. Cambridge University: Press.

Hausman, D. (2015). *Valuing health: Well-being, freedom, and suffering*. Oxford: Oxford University Press.

Hausman, D., & McPherson, M. (2007). The philosophical foundations of mainstream normative economics. In *The philosophy of economics: An anthology* (pp. 226–250). Cambridge: Cambridge University Press.

Karimi, M., Brazier, J., & Basarir, H. (2016). The capability approach: A critical review of its application in health economics. *Value in Health, 19*, 795–799.

Kaufman, A. (Ed.). (2005). *Capabilities equality: Basic issues and problems*. New York: Routledge.

Kellert, S., Longino, H. E., & Waters, K. (Eds.). (2006). *Scientific pluralism* (Minnesota studies in the philosophy of science) (Vol. XIX). Minneapolis: University of Minnesota Press.

Kincaid, H., Dupré, J., & Wylie, A. (Eds.). (2007). *Value-free science: Ideals and illusions*. Oxford: Oxford University Press.

Kitcher, P. (2001). *Science, truth and democracy*. Oxford: Oxford University Press.

Laudan, L. (1984). *Science and values*. Los Angeles-London: University of California Press.

Lawson, T. (1997). *Economics and reality*. London: Routledge.

Longino, H. E. (1990). *Science as social knowledge: Values and objectivity in scientific inquiry*. Princeton: Princeton University Press.

Longino, H. E. (2019). The *social dimensions of scientific knowledge*. In: E. N. Zalta (Ed.), *The stanford encyclopedia of philosophy*. Retrieved: https://plato.stanford.edu/entries/scientific-knowledge-social. Accessed 20 Aug 2020.

Machamer, P. K., & Wolters, G. (Eds.). (2004). *Science, values and objectivity*. Pittsburgh: Pittsburgh University Press.

Mäki, U. (2011). Scientific realism as a challenge to economics (and vice versa). *Journal of Economic Methodology, 18*(1), 1–12.

Mari-Klose, P. (2000). *Elección racional*. Madrid: Centro de Investigaciones Sociológicas.

McMullin, E. (1982). Values in science. *PSA: Proceedings of the Biennial Meeting of the Philosophy of Science Association, 4*, 3–28.

Mueller, D. (2003). *Public choice III*. Cambridge: Cambridge University Press.

Munda, G. (2016). Beyond welfare economics: Some methodological issues. *Journal of Economic Methodology, 23*(2), 185–202.

Nagel, T. (1986). *The view from nowhere*. Oxford: Oxford University Press.

Nozick, R. (1974). *Anarchy, state and utopia*. New York: Basic Books.

Nussbaum, M. (2000). *Women and human development: The capabilities approach*. Cambridge: Cambridge University Press.

Osmani, S. R. (2009). The Sen's system of social evaluation. In K. Basu & R. Kanbur (Eds.), *Arguments for a better world: Essays in honour of Amartya Sen* (Ethics, welfare and measurement) (Vol. 1, pp. 15–34). Oxford: Oxford University Press.

Osmani, S. R. (2010). Theory of justice for an imperfect world: Exploring Amartya Sen's idea of justice. *Journal of Human Development and Capabilities, 11*(4), 599–607.

Peter, F. (2012). Sen's idea of justice and the locus of normative reasoning. *Journal of Economic Methodology, 19*(2), 165–167.

Putnam, H. (2002). *The collapse of the fact/value dichotomy and other essays*. Cambridge Mass-London: Harvard University Press.

Putnam, H., & Walsh, V. (2012). *The end of value–free economics*. Abingdon: Routledge UK Publications.

Rawls, J. (1970). *The theory of justice*. Cambridge, MA: Harvard University Press.

Rawls, J. (1993). *Political liberalism*. New York: Columbia University Press.

Rawls, J. (1999). *The law of people*. Cambridge: Harvard University Press.

Reiss, J. (2013). *Philosophy of economics. A contemporary introduction*. New York: Routledge Contemporary Introduction to Philosophy.

Reiss, J. (2017). Fact-value entanglement in positive economics. *Journal of Economic Methodology, 24*(2), 134–149.

Rescher, N. (1988). *Rationality: A philosophical inquiry into the nature and the rationale of reason*. Oxford: Clarendon Press.

Rescher, N. (1997). *Objectivity: The obligations of impersonal reasons*. Notre Dame: Notre Dame University Press.

Robbins, L. ([1945] 2007). The nature and significance of economic science. Reprinted in: D. M. Hausman (Ed.), *The philosophy of economics: An anthology* (pp. 73–99). Cambridge: Cambridge University Press.

Scanlon, T. (1998). *What we owe to each other*. Cambridge: Harvard University Press.

Sen, A. (1969). Quasi-transitivity, rational choice and collective decisions. *Review of Economic Studies, 36*, 381–393.

Sen, A. (1970). *Collective choice and social welfare*. San Francisco: Holden-day.

Sen, A. (1977). Rational fools: A critique of the behavioural foundations of economic theory. *Philosophy and Public Affairs, 6*, 317–344.

Sen, A. (1979). Personal utilities and public judgements: Or what's wrong with welfare economics? *Economic Journal Royal Economic Society, 89*(355), 537–558.

Sen, A. (1980). Equality of what? *Tanner Lectures on Human Values, 1*, 195–220.

Sen, A. (1982). *Choice, welfare and measurement*. Cambridge, MA: The MIT Press.

Sen, A. (1985a). Well-being, agency and freedom: The Dewey lectures 1984. *The Journal of Philosophy, 82*(4), 169–221.

Sen, A. (1985b). Goal, commitments and identity. *Journal of Law, Economics and Organization, 1*(fall), 341–355.

Sen, A. (1986). Prediction and economic theory. *Proceedings Royal Society London A, 407*, 3–23.

Sen, A. (1987). *On ethics and economics*. Oxford: Basil Blackwell.

Sen, A. (1993a). Internal consistency of choice. *Econometrica, 61*(3), 495–521.

Sen, A. (1993b). Positional objectivity. *Philosophy and Public Affairs, 22*(2), 126–145.

Sen, A. (1994). The formulation of rational choice. *The American Economic Review, 84*(2), 385–390.

Sen, A. (1997). Maximization and the act of choice. *Econometrica, 65*, 745–779.

Sen, A. (1999a). *Development as freedom*. New York: Alfred Knopf Publishing.

Sen, A. (1999b). The possibility of social choice. *American Economic Review, 89*, 178–215.

Sen, A. (2000a). The discipline of cost-benefit analysis. *The Journal of Legal Studies, 29*(S2), 931–952.

Sen, A. (2000b). Consequential evaluation and practical reason. *Journal of Philosophy, 97*, 477–502.

Sen, A. (2002). *Rationality and freedom*. Cambridge: Harvard University Press.

Sen, A. (2004). Capabilities, lists, and public reason: Continuing the conversation. *Feminist Economics, 10*(3), 77–80.

Sen, A. (2005). Why exactly is commitment important for rationality? *Economics and Philosophy, 21*, 5–13.

Sen, A. (2009a). *The idea of justice*. Cambridge, MA: Belknap Press.

Sen, A. (2009b). Response. In R. Gotoh & P. Dumouchel (Eds.), *Against injustice: The new economics of Amartya Sen* (pp. 297–309). New York: Cambridge University Press.

Sen, A. (2012a). Values and justice. *Journal of Economic Methodology, 19*(2), 101–108.

Sen, A. (2012b). A reply to Robeyns, Peter and Davis. *Journal of Economic Methodology, 19*(2), 17317–17316.

Sen, A. (2017). Reasoning and justice: The maximal and the optimal. *Philosophy, 92*, 5–19.

Sen, A. (2018). The importance of incompleteness. *International Journal of Economic Theory, 14*, 9–20.

Sen, A., & Williams, B. (1982). *Utilitarianism and beyond*. Cambridge: Cambridge University Press.

Sent, E.-J. (2006). Pluralism in economics. In S. Kellert, H. E. Longino, & K. Waters (Eds.), *Scientific pluralism* (Minnesota studies in the philosophy of science) (Vol. XIX, pp. 80–101). Minneapolis: University of Minnesota Press.

Van Staveren, I. (2007). Beyond utilitarianism and deontology: Ethics in economics. *Review of Political Economy, 19*(1), 21–35.

Vizard, P. (2005). *The contribution of professor Amartya Sen to human rights* (CASE paper 91). London: Centre for the Analysis of Social Exclusion (LSE).

Weber, M. ([1904] 2007). Objectivity and understanding in economics. Reprinted in: D. M. Hausman (Ed.), *The philosophy of economics: An anthology* (pp. 59–72). Cambridge: Cambridge University Press.

Write, J. (2018). Rescuing objectivity: A contextualist proposal. *Philosophy of the Social Sciences, 48*, 1–22. https://doi.org/10.1177/0048393118767089.

Index of Names

© Springer Nature Switzerland AG 2020
W. J. Gonzalez (ed.), *Methodological Prospects for Scientific Research*, Synthese
Library 430, https://doi.org/10.1007/978-3-030-52500-2

235

Subject Index

Printed in the United States
by Baker & Taylor Publisher Services